On the Study of Human Cooperation via Computer Simulation

Why Existing Computer Models Fail to Tell Us Much of Anything

Synthesis Lectures on Games and Computational Intelligence

Editor
Daniel Ashlock, *University of Guelph*

Synthesis Lectures on Games & Computational Intelligence is an innovative resource consisting of 75-150 page books on topics pertaining to digital games, including game playing and game solving algorithms; game design techniques; artificial and computational intelligence techniques for game design, play, and analysis; classical game theory in a digital environment, and automatic content generation for games. The scope includes the topics relevant to conferences like IEEE-CIG, AAAI-AIIDE, DIGRA, and FDG conferences as well as the games special sessions of the WCCI and GECCO conferences.

On the Study of Human Cooperation via Computer Simulation:
Why Existing Computer Models Fail to Tell Us Much of Anything
Garrison W. Greenwood

ISBN: 978-3-031-00993-8   paperback
ISBN: 978-3-031-02121-3   ebook
ISBN: 978-3-031-00170-3   hardcover

DOI 10.1007/978-3-031-02121-3

A Publication in the Springer series
SYNTHESIS LECTURES ON GAMES AND COMPUTATIONAL INTELLIGENCE

Lecture #4
Series Editor: Daniel Ashlock, *University of Guelph*
Series ISSN
Print 2573-6485   Electronic 2573-6493

On the Study of
Human Cooperation
via Computer Simulation

Why Existing Computer Models Fail to
Tell Us Much of Anything

Garrison W. Greenwood
Portland State University

SYNTHESIS LECTURES ON GAMES AND COMPUTATIONAL INTELLIGENCE #4

ABSTRACT

Cooperation is pervasive throughout nature, but its origin remains an open question. For decades, social scientists, business leaders, and economists have struggled with an important question: why is cooperation so ubiquitous among unrelated humans? The answers would have profound effects because anything that promotes cooperation leads to more productive work environments and benefits society at large. Game theory provides an ideal framework for studying social dilemmas, or those situations in which people decide whether to cooperate with others (benefitting the group) or defect by prioritizing their self-interest (benefitting only the individual). The social dilemma is formulated as a mathematical game and then programmed into a computer model. Simulating the game allows researchers to investigate potential theories to explain how cooperation emerges and what promotes its persistence.

Over the past 25 years, countless papers on social dilemma games have been published, yet arguably little progress has been made. The problem is the social dilemma game models are unrealistic in the sense they contain artificial constructs that deviate from the way humans act. This book describes the shortcomings in current social dilemma game modeling techniques and provides guidance on designing more effective models. A basic introduction to game theory is provided with an emphasis on the prisoner's dilemma, the most widely studied social dilemma game. Individual chapters are provided detailing the shortcomings of weak selection, spatial games, and the Moran process. Computer model validation is also discussed at length. The recommendations found in this book should help design more realistic social dilemma game models likely to produce a better understanding of human cooperation.

KEYWORDS

human cooperation, game models, social dilemmas, prisoner's dilemma, spatial games, Moran process, economic games

Contents

Preface

The most allusive, and potentially significant, open question in nature involves the genesis of cooperation. Cooperation is ubiquitous; it is found throughout the animal kingdom and even in the plant world. Among humans, cooperation with kin, such as an uncle or sister, is nothing unusual and might even be expected. But one question has plagued economists, business leaders, and social scientists for decades: why is cooperation so prevalent among unrelated humans?

The answer to that question would have profound effects. Anything that promotes mutual cooperation improves society. Companies, knowing the conditions that promote cooperation, could create more productive work environments. Inner-city communities could offer cleaner neighborhoods with less crime. Mutual cooperation leads to more effective and less costly use of public resources.

Imagine a group of N people who are all engaged in some kind of economic activity. This activity is repeated for a finite (but not announced) number of rounds. In each round every individual interacts with one or more other individuals and receives a payoff. The payoff size depends not only on the choices the individual makes, but also on the choices made by the individuals he interacts with. Each individual faces a dilemma: should he be self-interested and make choices that benefit only himself or make choices that benefit the group at large? In its simplest form, the individual chooses to either cooperate or defect. Cooperation benefits the group while defection benefits the individual.

These economic activities are collectively called social dilemmas. There are many real-world examples of social dilemmas. For example, charities rely on contributions. Cooperators contribute while defectors do not. Everyone can potential benefit from the services provided by the charity whether or not they contribute. But if nobody contributed the charity would quickly go out of business. Taxes pay for city parks and libraries. Everyone can use these facilities although not everyone pays taxes. Without any tax revenue these facilities would close.

All social dilemmas have two conflicting properties: (1) defectors always do better than cooperators regardless of what actions others might take; and (2) mutual cooperation is the best group outcome. Indeed, it turns out mutual cooperation always has higher payoffs to all participants than mutual defection. Unfortunately, the first property makes defection quite tempting which makes mutual defection a likely outcome in all real-world social dilemmas. Everybody loses if everybody defects.

Game theory provides an ideal environment for studying social dilemmas. The social dilemma is formulated as a mathematical game which is then programmed to create a computer model. This computer model gives researchers something to play with. Different strategies can be tested and different circumstances that promote cooperation in human populations can be investigated. Prisoner's dilemma (PD) is the most widely known social dilemma game while the most widely studied is the public goods game (PGG), which the N-player version of PD ($N > 2$). Researchers posit some strategy or other conditions that should promote cooperation. The hypothesis is tested by creating a computer model for the game and then simulations are run to predict how cooperation levels in a population change over time. The hypothesis is confirmed if the cooperation levels change as predicted. However, final confirmation requires validation that compares model predictions against data from human experiments.

Over the last 25 years or so, literally thousands of papers have been published regarding cooperation in social dilemmas. A number of ideas have surfaced such as reciprocity, altruistic punishment, and so forth. Yet, despite all of the papers published, despite all of the computer models constructed and simulated, arguably our understanding of cooperation among non-kin has made little progress. We don't really know all that much more than we did 25 years ago. Why not? This is a puzzling—and somewhat disappointing—situation.

A survey of the literature reveals some reasons why progress has stalled. There are several recurring techniques incorporated into the computer models. For instance, even in large populations, humans tend to associate with relatively few other individuals. Spatial and network games are used to address local interactions. Weak selection is often used to linearize mathematical equations, which makes model analysis easier. The Moran process is extensively used to evolve strategies in the population. Therein lies the problem: these computer modeling techniques inadvertently make the models incapable of getting the answers. These techniques are artificial constructs that considerably deviate from the way humans actually behave in social dilemmas. Consequently, it is highly unlikely that models that incorporate have these techniques will be able to provide any insight into human cooperation.

Progress in understanding cooperation will continue to languish unless radical changes are made in the way social dilemma computer models are designed. The purpose of this book is to describe the social dilemma computer models in current use and to describe their inherent shortcomings. These shortcomings are not isolated cases but exist in an overwhelming majority of current computer models described in the literature. It is hoped that the recommendations provided will help create more realistic and effective social dilemma computer models in the future.

The book is organized as follows. Chapter 1 provides formal definitions of a social dilemma and describes several real-world examples. The three most widely studied social dilemma games are described in Chapter 2. Spatial games and network games are closely related and are discussed in Chapter 3. The problems with weak selection are described in Chapter 4. The Moran process and replicator equations are the two primary techniques used to evolve strategy frequencies in populations. Chapter 5 describes these two methods. Every computer model tests a hypothesis about cooperation and makes predictions about how cooperation levels will change under that hypothesis. Unfortunately, most models contain unrealistic features (such as enormous population sizes). These features make it virtually impossible to validate the model's predictions. Finally, Chapter 7 discusses the difficulty of validating the current computer models. Some recommendations on possible replacements for the Moran process and spatial games are presented.

Garrison W. Greenwood
July 2019

Acknowledgments

To my wife, Linda, and my two children, Matthew and Sarah, whose encouragement made this book possible.

Garrison W. Greenwood
July 2019

CHAPTER 1

Social Dilemmas

Social dilemmas exist whenever a tradeoff must be made between what is best for an individual in a group and what is best for the group overall. Interest in social dilemmas has exploded over the last few decades because they describe pressing social problems such as water supply usage, energy consumption, funding government programs, and pollution control.

In social dilemmas individuals choose whether to cooperate or to defect and receive a payoff or reward based on their choice and the choices made by others in the group. Cooperators make group interests paramount while defectors act solely in their own self-interest. It is worth noting that pursuing one's self-interest is not necessarily a bad thing; the famous Scottish economist Adam Smith stated *"In competition, individual ambition serves the common good."* Nevertheless, in social dilemmas there is always some contention between cooperation and defection because defection by an individual benefits only that individual. Put another way, in social dilemmas, unlike the economy at large, acting in one's own self interest is often perceived as selfish and without regard for group interests. Dawes [1980] identified two seemingly conflicting properties all social dilemmas have: (1) an individual gets a higher payoff by defecting, regardless of what others do, but (2) the payoff is lower if all individuals defect than if all individuals cooperate.

Those two properties explain why social dilemmas are so important. Whenever a cooperator interacts with a defector, the defector will always do better (get a higher reward). Consequently, there is a strong temptation to defect instead of cooperate. But if everyone defects, then the whole group loses. Unfortunately, mutual defection is a likely outcome of all real-world social dilemmas—even though mutual cooperation produces the highest reward for a group. This contradiction is the reason why social dilemmas have attracted so much interest. If the whole group benefits the most when everybody cooperates, why does universal defection prevail? Under what conditions could cooperation persist or hopefully even grow? The answers to these questions would have profound positive effects on society. Yet, currently, these are still open questions.

A few definitions are needed before giving some social dilemma examples. An *economic good* is anything that can be used or consumed by society. Economic goods can be classified by two characteristics [Dionisio and Gordo, 2006].

Definition 1.1 (**excludable good**) *A good is excludable if not everyone in a group can use or consume it.*

Definition 1.2 (**diminishable good**) *A good is diminishable if the use or consumption by one individual reduces the use or consumption by another individual.*

Being excludable or not is independent of being diminishable or not. That is, these are two independent characteristics.

Cost can make a good excludable because not everyone may be able to afford it. Cable television is an example. But there are other reasons that could make a good excludable. For example, toothbrushes or deodorant sticks are inexpensive and readily available but are used by only one individual for hygienic reasons.

A diminishable good is, by definition, a finite resource so its consumption or use is limited. Consider rush hour traffic on a public highway. The highway is clearly a diminishable good because cars currently on the highway prevent other cars from using the highway at the same time. Since it is a public highway it is non-excludable. Public land such as city parks are another example. Every deer a hunter shoots is a deer not available to another hunter. Deer hunting is diminishable but also excludable because only those who can pay for a hunting license can participate. Tax revenues pay for national defense. Everyone in society benefits from national defense whether they pay taxes or not. Consequently, national defense is an example of a non-excludable good. It is also non-diminishable. Non-excludable and non-diminishable goods are called *public goods*.

Social dilemmas are generally non-excludable; everyone can participate either voluntarily or by default. Funding for national public radio (NPR) is a social dilemma with voluntary participation. A significant portion of their annual budget comes through private donations. But anyone can listen to the broadcasts whether or not they contributed anything. Of course, if nobody contributed then this would negatively affect the variety and quality of the shows. It is non-exclusionary in the sense anybody who listens can donate. Now consider a factory owner who has to decide whether she will limit air pollution from the factory. There is a cost associated with cooperation making it less profitable than defection. However, society at large pays a price (health, etc.) if air pollution is not limited. Thus, everyone participates in this social dilemma by default.

Non-excludable social dilemmas have to contend with the *free rider problem*. Free riders are individuals who consume a public good but contribute nothing to it. People who receive protection afforded by national defense, but pay no taxes, are free riders. Too many free riders can deplete a public good even though, by definition, it should not be diminishable. National defense is naturally non-diminishable—providing protection to one individual does not limit protection to another—but if nobody paid taxes there would be no money to provide protection to anybody. This situation is unlikely to ever occur so it is not considered. Free riders are a concern, however, because they take advantage of those who are cooperating (albeit not as much as defectors). Free riding causes resentment and can often result in retaliatory actions.

There are two classes of social dilemmas depending on whether they involve diminishable or non-dimishable resources. The NPR example is non-diminishable because one person listening to a broadcast does not prevent someone else from listening. Diminishable resource social dilemmas follow a *tragedy of the commons* paradigm which was proposed by Hardin [1968]. Imagine a commons area where shepherds can graze their sheep. The commons can only accommodate a certain number of sheep otherwise overgrazing will eventually deplete the grass. A cooperator would limit the size of his herd and the time his sheep can graze. A defector contemplates if he can add more sheep to his herd, thereby improving his profit. There is a strong temptation to defect because over-consumption by some shepherds exploits those who don't and nobody likes to be exploited. However, if everyone defected eventually the commons would be unusable. The grass in the commons is diminishable. Competition for diminishable goods often leads to *rivalry* among individuals.

The inevitable outcome of all social dilemmas is mutual defection whenever cooperation or defection are the only strategy choices. Research in this area has one objective: identify conditions that "resolve" the dilemma. A social dilemma is considered resolved if cooperation grows (or at least persists) in a population over time thereby avoiding the mutual defection state.

Nowak [2006] posited five rules that could promote cooperation in populations: direct reciprocity, indirect reciprocity, network reciprocity, kin selection, and group selection. Direct reciprocity means one individual decides to cooperate or defect with another individual depending on their past interactions. That is, an individual is more likely to cooperate with someone in the future if they cooperated in previous encounters. Indirect reciprocity is based on reputation. Even though two individuals have had no previous interactions, one is likely to cooperate with someone who has a history of cooperating with others. Well mixed populations are populations where every individual can interact with every other individual with some probability. In the real world at large, well-mixed populations are rare; people tend to interact over and over with only a subset of the population. Network reciprocity captures this idea by assuming there is some underlying spatial structure in the population. Individuals can be thought of as residing on the vertices in a graph and the edges between vertices indicate the interactions. Kin selection says individuals are more likely to cooperate with those they are genetically related to (an uncle or sister for example). Finally, group selection says small groups of cooperators that repeatedly interact may persist in a population of mostly defectors.

The problem with Nowak's rules is they deal primarily with pairwise interactions making them practically too simplistic to explain cooperation in large groups. For instance, after 2 or 3 interactions indirect reciprocity becomes meaningless because an individual now has real data on how another behaves. It is unclear how direct reciprocity can help make cooperate/defect decisions in groups where say half of the group cooperated and the other half defected. His rules also do not consider deliberate strategy choices an individual might make to influence future strategy choices by others.

A number of theories have proposed ways to resolve social dilemmas in large groups. One theory promotes the idea of *altruistic punishment*. The idea behind punishment is defectors are levied a penalty, such as a fine, which reduces the payoff for defecting. This lower payoff should deter future defection. The punishment is usually costly in the sense punishers pay some small cost—e.g., money or effort—to inflict punishment upon a defector. Costly punishment has been observed in many diverse societies (see Henrich et al. [2006]).

Punishers are also cooperators although their payoff is slightly less because punishment is costly. But cooperators who do not punish are free riders because they benefit from the punishment imposed on defectors while letting others pay the associated cost. Defectors are called *first-order free riders* whereas cooperators who don't punish are called *second-order free riders*. Punishing both first- and second-order free riders to prevent being exploited is always a possibility.

Fehr and Bächter [2002] defined two components of altruistic punishment: (1) there is no expectation of future material gain; any rewards are conferred on others; and (2) there is a cost that is paid to inflict the punishment on others. This definition is widely accepted. However, any belief that the first component is a necessary ingredient of punishment in social dilemmas is somewhat naive. As pointed out by Greenwood et al. [2018], the whole idea behind punishment is to get defectors to stop defecting. If successful, everyone benefits—including the punishers—and everybody knows that a priori. Punishers therefore punish on purpose and expect to benefit from the anticipated increase in cooperation levels. While punishment can help resolve a social dilemma, it is motivated more by self interest than altruism.

Emotions can influence strategy changes in social dilemmas. Seip et al. [2014] conducted human experiments and concluded anger can motivate cooperators to engage in costly punishment. It is not the event itself that causes the strategy change but a perception of unfair treatment by others. Greenwood [2015b] suggested guilt can play a role in resolving a tragedy of the commons. Consider, for example, commercial fisherman in the Tasman Sea off the coast of Australia. Defectors will try to maximize their profit by taking as many fish as possible whereas cooperators will limit their catch to give the fish a chance to repopulate. Defectors may eventually realize the fish population is becoming depleted and decide to also limit how many fish they take.

Games involve competition among individuals. Mathematical games provide any ideal platform for investigating how and why cooperation persists in social dilemmas. That aspect is explored in the next chapter.

CHAPTER 2

Prisoner's Dilemma and Other Social Dilemma Games

This chapter begins with an overview of game theory. It is not intended to comprehensively look at the entire game theory field but does provide sufficient detail to understand how social dilemma games work. Indeed, the definitions are given in the context of social dilemma games. This overview is followed by a description of the three most widely studied games used in social dilemma investigations: the prison's dilemma game, the public goods game (which is an $N > 2$ player version of prisoner's dilemma), and the snowdrift game.

2.1 GAME THEORY

Games are competitions between two or more individuals. Game theory tries to discover why players make particular choices during the game. In one-shot games players interact once and the game is over. Iterated games repeat for a finite number of rounds, which allows history from previous rounds to influence future game moves. Although some researchers study one-shot social dilemma games, the vast majority focus on iterated games.

Players alternate moves in sequential games. A game tree is used to model such games. Chess is an example of a sequential game. On the other hand, social dilemma games are simultaneous games; players announce their moves at the same time and they receive a reward or payoff depending on their move and the moves others make. For 2-player games these rewards are usually represented by a payoff bi-matrix.

Figure 2.1 shows an example of a payoff bi-matrix. Here "A" and "B" represent player's moves in a round of the game. If, for instance, the row player chooses move "A" and the column player move "B", then they receive a reward of 0 units and 1 unit, respectively.

Remark 2.1 If the game has a symmetric bi-matrix—i.e., the game doesn't change if the two players switch roles—then normally only a single payoff matrix is used showing the payoffs for just the row player.

payoff to row player   payoff to column player

$$
\begin{array}{cc}
 & A \quad\quad B \\
\begin{array}{c} A \\ B \end{array} &
\left(\begin{array}{cc}
2,2 & 0,1 \\
1,0 & 4,-1
\end{array} \right)
\end{array}
$$

Figure 2.1: A typical game payoff bi-matrix.

Definition 2.2 (**normal form game**) *An N -player normal form game consists of*

- *a set players indexed by i* $\in \{1, 2, 3, \ldots, N\}$

- *a set* A_i *of actions player i can choose from*

- *a utility function* $u_i : A_1 \times \cdots \times A_N \to \mathbb{R}$.

Social dilemma games are normal form games. A player's *utility* (also known as a payoff or reward) depends not only on his choice but the choices of all $N - 1$ other players. This utility could be a monetary reward, indication of a win or loss, of just information about the new game state. Moving a pawn in a chess game may not end the game but it does give information about the new game state. Utility is almost always a monetary reward in social dilemma games. A player is considered *rational* if he tries to maximize the expected value of u_i.

An *action* is a player's next move in a game (such as cooperate or defect), while a *strategy* is the underlying thought process behind picking that particular action. In other words, actions implement strategies. It is common to see researchers conflate these two terms. For example, "defection" is a strategy for playing the game and at the same time an action that produces a payoff say from a payoff matrix. The explanation for this conflation is switching from say cooperation to defection is an action change, but any action change must be caused by a strategy change. Therefore, strategy changes and action changes are really the same thing.

Remark 2.3   An example where this explanation falls apart is the Tit-For-Tat strategy used in the 2-player iterated prisoner's dilemma game. This strategy can choose cooperate in the next round and defect in the following round depending on what action the other player takes. Two different actions, but the same strategy. Nevertheless, conflating strategy with action is usually not problematic because the context is quite clear.

In social dilemma games the action choices available are the same for every player—i.e., $A_i = A_j \ \forall i, j$. However, in any given round one player does not know what specific strategy another player has chosen beforehand. The payoff each player receives after each round is disclosed, but a player can only infer what strategy the other player is using. Post analysis of a social dilemma game tries to identify why players cooperated or defected and under what circumstances cooperation levels will persist.

Definition 2.4 (**Nash equilibrium**) *A Nash equilibrium (NE) is a selection of strategies where no player can unilaterally switch to a different strategy and receive a higher payoff.*

This definition can be expressed more formally. Let $s_i \in S$ be a strategy from a strategy set S and denote by s_{-i} all strategies in S except s_i. Let $\pi(s_i, s_j)$ denote the payoff to a player using strategy s_i when competing against a player using strategy s_j. s_i is a "best response" to s_{-i} if and only if

$$\pi(s_i, s_{-i}) \geq \pi(s_i', s_{-i}) \quad \forall s_i' \in S. \tag{2.1}$$

A selection of strategies is then a NE if, for each player i, his strategy s_i is a best response to s_{-i}.

Definition 2.5 (**Pareto optimal**) *A selection of strategies is Pareto optimal (PO) if any unilateral strategy change that results in a higher payoff for one player will result in a lower payoff for some other player.*

A NE can be PO, but not always. For instance, in the payoff bi-matrix

$$\begin{array}{c} & \begin{array}{cc} A & B \end{array} \\ \begin{array}{c} A \\ B \end{array} & \left(\begin{array}{cc} 2,2 & -1,3 \\ 3,-1 & 0,0 \end{array} \right) \end{array} \tag{2.2}$$

the player strategy pair (A, A) is PO but (B, B) is the only NE. All players might do better leaving a NE that is not also PO.

The NE is particularly important in social dilemma games. Most of the really interesting work investigates how strategies evolve over time in N-player populations ($N \gg 2$). Nash equilibria are fixed points where no further evolution takes place.

Not all games have unique NE. This led researchers to search for equilibrium "refinements." These refinements are a set of rules that would identify and exclude sub-optimal NE. Unfortunately these efforts were frustrated because refinements that worked for one game wouldn't always work for other games. This frustration led to the field of *evolutionary game theory* where strategy choices in a population evolve over time eventually converging at the optimal NE. Players now adapt their behavior as the game is repeatedly played. This brings beliefs in line with behavior (a NE requirement). Equilibrium can now be viewed as an adjustment process rather than something that just happens.

Let $S = (s_1, \ldots, s_n)$ be a strategy profile. Each s_i is a unique, complete definition of how to play the game and is called a *pure strategy*. A play may choose a pure strategy to play the game or decide to make a random choice among the n pure strategies. This is called a *mixed strategy*. Assume there is some probability x_i of choosing strategy $s_i \in S$. This gives a mixed strategy profile $x = (x_1, \ldots, x_n)$ where $\forall i$, $x_i \geq 0$ and $\sum_i x_i = 1$. Suppose there are two players with mixed strategy profiles x and y. Then with payoff matrix W, the expected payoff is

$$\pi(x, y) = xWy^{\mathrm{T}}. \tag{2.3}$$

Example 2.6 Consider a game with two players and two pure strategies cooperate (C) and defect (D). The players have strategy profiles $x = (1/2, 1/2)$ and $y = (2/3, 1/3)$ and the payoff matrix is

$$W = \begin{array}{c} \\ C \\ D \end{array} \begin{array}{c} C \quad D \\ \begin{pmatrix} 3 & 0 \\ 5 & 1 \end{pmatrix} \end{array}.$$

Then the expected payoff is

$$\pi(x, y) = xWy^{\mathrm{T}} = \frac{17}{6}.$$

Smith [1974] introduced the idea of an *evolutionary stable strategy* (ESS). Consider a finite size population that plays multiple rounds of a game. Let σ be the current strategy used by the population and let τ be an alternative strategy that gets introduced into the population via mutation. Only a very small fraction of the population ($< 1\%$) become mutants. The population is stable if it cannot be invaded (taken over) by the mutants. The mutants may initially grow in frequency and perhaps even persist for a short period of time. But if σ is an ESS, eventually the mutants will go extinct leaving σ as the only strategy in the population.

σ is an ESS if and only if the following holds:

- $\pi(\sigma, \sigma) > \pi(\tau, \sigma)$ (i.e., σ is better against σ than τ is against σ).

- or, if $\pi(\sigma, \sigma) = \pi(\tau, \sigma)$ then $\pi(\sigma, \tau) > \pi(\tau, \tau)$ (i.e., if σ and τ are equally good against σ, then σ is better against τ than τ is against another τ).

Example 2.7 (Hawk-Dove game)
A 2-player, two-strategy game has a payoff bi-matrix

$$W = \begin{array}{c} \\ \text{hawk} \\ \text{dove} \end{array} \begin{array}{c} \text{hawk} \quad \text{dove} \\ \begin{pmatrix} -1, -1 & 2, 0 \\ 0, 2 & 1, 1 \end{pmatrix} \end{array}.$$

The population is nearly homogeneous, consisting almost entirely of doves (σ) but has a tiny mutant sub-population of hawks (τ). Then the following observations can be made.

- $\pi(\sigma, \sigma) = 1 < 2 = \pi(\tau, \sigma)$ so hawks will grow in the population. (They can invade.)

- But $\pi(\sigma, \tau) = 0 > -1 = \pi(\tau, \tau)$ so the number of hawks will grow only so long as the population consists mostly of doves

- Hawks cannot completely take over the population

- The mixed strategy $(D, H) = \left(\frac{1}{2}, \frac{1}{2}\right)$ is an ESS. It is also a NE if both players use it.

There are two approaches to EGT. The first approach uses Maynard Smith's ESS work while the second approach creates a model that uses some evolutionary process. The first approach only defines stability while ignoring the process whereby populations evolve; it only concentrates on finding the long-term behavior. The second approach does not define stability. Instead, it explicitly models the dynamics of an evolutionary process and sees what happens to the strategy frequencies in the population over time. The vast majority of social dilemma game researchers use the second, model-based approach. However, there are some serious questions about the validity and credibility of many of the models that have appeared in the literature. These concerns will be discussed at depth in later chapters.

2.2   THE PRISONER'S DILEMMA GAME

The original game was developed by Merrill Flood and Melvin Dresher at the Rand Corporation in 1950. It was later formalized as a prisoner's dilemma game by Albert Tucker (who incidentally was John Nash's faculty adviser at Princeton). The game was introduced by the following story.

> *Two suspects are incarcerated by the police and held in separate interrogation rooms so they cannot communicate with each other. They are encouraged to confess and implicate the other. Each is informed of the following consequences:*
>
> — *if each of you betrays the other, you both get 2 years in prison*
> — *if you betray your partner, and she remains silent, you go free and she gets 5 years in prison*
> — *if both of you remain silent, there is still enough evidence to convict both of you of a lesser crime. You will both serve one year in prison.*

A suspect cooperates (C) with her partner by remaining silent but defects (D) by betraying her. There are four possible outcomes. (1) You defect and your partner cooperates (you go free, 5 years for her). (2) You both cooperate (1 year each). (3) You both defect (2 years each). (4) You cooperate, and she defects (5 years for you, she goes free). The payoff bi-matrix is

$$
\begin{array}{cc}
 & \begin{array}{cc} C & \quad D \end{array} \\
\begin{array}{c} C \\ D \end{array} & \left(\begin{array}{cc} -1, -1 & -5, 0 \\ 0, -5 & -2, -2 \end{array} \right).
\end{array}
$$

It is common practice to represent the payoff matrix as

$$\begin{array}{cc} & \begin{array}{cc} C & D \end{array} \\ \begin{array}{c} \text{payoff to } C \\ \text{payoff to } D \end{array} & \begin{pmatrix} R & S \\ T & P \end{pmatrix} \end{array} \tag{2.4}$$

or in bi-matrix form as

$$\begin{array}{cc} & \begin{array}{cc} C & D \end{array} \\ \begin{array}{c} C \\ D \end{array} & \begin{pmatrix} R,R & S,T \\ T,S & P,P \end{pmatrix}, \end{array} \tag{2.5}$$

where T is the temptation to defect, R is the reward for cooperating, S is the sucker's payoff, and P is for punishment both players receive for defecting. To create a prisoner's dilemma it is necessary that

$$S < P < R < T.$$

Defection seems to be the best choice regardless of what the other suspect does. Therein lies the dilemma: rational players wind up getting P instead of the higher payoff R resulting from mutual cooperation. The strategy pair (D, D) is a NE while (C, C) is PO.

In the iterated prisoner's dilemma (IPD) game there is a constraint

$$2R > T + S \tag{2.6}$$

to prevent player collusion by alternating between cooperation and defection thereby getting a greater reward than mutual cooperation.

Remark 2.8   The values $T = 5$, $R = 3$, $P = 1$, and $S = 0$ appear frequently in the literature.

The PD payoff matrix can be reduced to an equivalent form but with only two independent parameters. This equivalent form has a slightly different interpretation of cooperation and defection, but maintains the relative payoff orderings required to produce a prisoner's dilemma. In this alternative form the payoff matrix is now expressed in terms of costs and benefits. A cooperator pays a cost c to provide some benefit b to the other player (where $b > c > 0$). This alternative payoff matrix is

$$\begin{array}{cc} & \begin{array}{cc} C & D \end{array} \\ \begin{array}{c} \text{payoff to } C \\ \text{payoff to } D \end{array} & \begin{pmatrix} b-c & -c \\ b & 0 \end{pmatrix}. \end{array} \tag{2.7}$$

There are over 100 known IPD strategies. The best known is *Tit-for-Tat* (TFT), which initially cooperates and thereafter does whatever the opponent did in the previous round. There have been a number of IPD tournaments as well (see Jurisic et al. [2012], Kendall et al. [2007]).

2.3 THE PUBLIC GOODS GAME

The public goods game (PGG) is an N-player version of IPD (where $N > 2$). It uses a well-mixed population. The game can be formulated in various ways and in this section two such forms are given. The first form does not use a payoff matrix. Each round is played as follows.

N players have an option of putting from $0–100 into an investment fund. They are told the fund will be multiplied by a factor r (where $1 < r < N$) and then split equally among all players. Investors must subtract their contributions from their returns.

Rational players exploit those who do contribute; they contribute nothing but get the same return as everybody else. The best group outcome occurs when everyone invests the maximum amount. However, investors always have a disadvantage.

Example 2.9 Four players each have an opportunity to put from $1 up to $10 into a pot. Players announce their decisions simultaneously. All players know the amount in the pot will be doubled and then re-distributed equally. Players who put money into the pot only get $0.50 back for every dollar contributed.

Cooperators contribute while defectors do not. Hence, defectors always do better. In the simplest form of the game cooperators contribute a fixed amount λ. In that case, the payoff to a defector is $P_D = r\lambda n_c/N$ and to a cooperator $P_C = P_D - \lambda$ where n_c is the number of cooperators in the group. Doebeli and Hauert [2005] describe a transformation of b and c in terms of r, λ, and N that makes the PD payoff matrix of Eq. (2.4) suited for use in PGGs. With this transformation a PGG with group size N is equivalent to $(N-1)$ pairwise IPD games.

2.4 THE SNOWDRIFT GAME

In IPD games defectors always prevail if cooperation and defection are the only strategies. The iterated snowdrift game (ISD) is different because it does allow some level of cooperation to remain in a majority population of defectors. The reason is the payoffs are adjusted so a cooperator still derives some benefit even if his opponent defects. Human experiments confirm some level of cooperation does persist in an ISD population (Kümmerli et al. [2007]). A two-player version of the game is described by the following scenario.

Two travelers get stuck in a snowbank. They can cooperate by shoveling snow to free both cars or they can defect by not shoveling the snow. They remain stuck if both sit in their cars but are freed if at least one of them shovels the snow.

Let b be the benefit of shoveling snow and c a cost representing the associated effort where $b > c$. This cost reduces to $c/2$ if they both shovel snow because the effort is shared. The snowdrift game payoff matrix is

$$
\begin{array}{cc}
 & \begin{array}{cc} C & \quad D \end{array} \\
\begin{array}{c} \text{payoff to } C \\ \text{payoff to } D \end{array} & \left(\begin{array}{cc} b - c/2 & b - c \\ b & 0 \end{array} \right).
\end{array}
\tag{2.8}
$$

Equation (2.4) can also represent the payoff matrix except now T is the temptation to not shovel snow, R is the reward if both drivers shovel snow, S is still a sucker who shoveled snow while the other driver stayed warm in his car, and P is the punishment for neither driver shoveling snow (they both remain stuck). This interpretation causes a reordering of the payoffs to $T > R > S > P$.

CHAPTER 3

Spatial and Network Games

This book focuses on iterated games with population size $N > 2$. It is therefore important to understand the population structure because that determines which players interact each round.

Some games have a *well-mixed population* where players interact with other players with some probability p. This probability is fixed and depends on the game. For instance, in some games, $p = 1$ because every player interacts with every other player per round. It is common practice, though, to subdivide the population into equal size groups and players interact only with group members; $p = 1$ for group members and $p = 0$ from those players outside the group. Other games have an individuals interact with only one other randomly chosen player in the population each round. In that case $p = 1/N$.

In large populations people tend to associate with only a few people. You may work for a company with 100 employees. The company makes and sells a single product. The goal is to have everyone cooperate so the company produces a high demand, affordable product. Yet you many only interact with five or six people in the company on a regular basis. Even in societies people tend to have more restricted, localized interactions due to cultural or ethnic norms. This tendency is mimicked in social dilemma games through spatial models and network models where players only interact with a fixed, finite number of other players. Interestingly enough, computer simulations showed cooperation can persist in such environments because repeated interactions with the same partners creates clusters of cooperators. Cooperation in many groups yields high cooperation levels in the entire population.

3.1 SPATIAL GAMES

Neighborhoods in social games contain players that repeatedly interact. Two players who belong to the same neighborhood are said to be neighbors. The size of a neighborhood depends on the game model. In spatial games players are uniquely assigned to points in a 2D square lattice. Neighbors are determined by Euclidean distance and all neighborhoods are the same size. The two most common neighborhoods are the *Von Neumann* and the *Moore* which are both shown in Figure 3.1. The Von Neumann type has neighbors to the north, south, east, and west whereas the Moore type adds neighbors to the northwest, southwest, southeast and northeast. Periodic boundary conditions are enforced to keep neighborhoods the same size. These boundary conditions provide a "wrap around" connection between the top and bottom rows and the left and right columns to make all neighborhoods the same. For instance, consider a focal player on the top row of a lattice with Von Neumann neighborhoods. Three neighbors are to the west (left

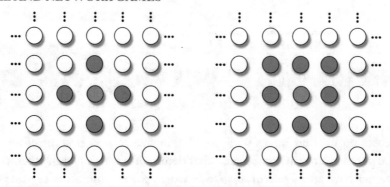

Figure 3.1: Spatial game showing Von Neumann (left) and Moore (right) neighborhoods. The focal player is shown in red. Each round the focal player only interacts with players in his neighborhood. Lattice has periodic boundary conditions (see text).

same row), east (right same row), and south (below same column). The fourth neighbor to the north is in the bottom row and in the same column.

3.2 NETWORK GAMES

Players in games with networks are assigned to vertices in an undirected graph. Two players are in the same neighborhood if they are connected by an edge. The number of neighbors a player has depends on the vertex degree—i.e., the number of edges incident to the vertex. Social games use regular, small-world, and scale-free networks but regular graphs are by far more likely to be used.

Regular graphs have a motif which makes all neighborhoods the same size. In other words, each vertex in the graph is incident to the same number of edges. An example is the 2D grid (with periodic boundary conditions) shown in Figure 3.2.

Almost all work with network games has used a 2D grid but some work has looked at two other kind of networks: small-world networks and scale-free networks. Small-world graphs are constructed by making random permutations to a regular graph. An example of this construction process is shown in Figure 3.3. On the left side is a regular graph (every vertex has the same degree, in this case 4). On the right side shows the resultant small-world network created by randomly moving two edges which are highlighted in red. One property of small-world networks is the number of hops from one vertex to any other vertex is less than (or equal to) the number of hops in the original regular graph. A method quite frequently used to create a small-world graph is to start with a 2D grid and Moore neighborhoods. Then a small number of edges are randomly chosen and rewired to connect to a vertex somewhere else in the grid.

Scale-free networks have a few vertices with high degree, called hubs, while the vast majority of vertices have a very small degree. A scale-free network has a power law degree distri-

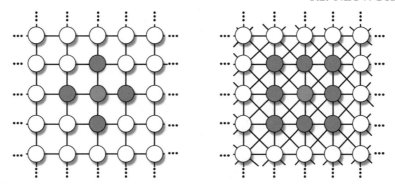

Figure 3.2: 2D grid networks showing Von Neumann (left) and Moore (right) neighborhoods. The focal is shown in red. The grid has periodic boundary conditions.

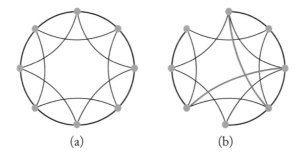

(a)   (b)

Figure 3.3: Regular graph (a) and conversion to a small-world graph (b).

bution

$$P_{\text{deg}}(k) \sim k^{-\gamma},$$

where k is the degree of a vertex and γ is a constant usually between 2 and 3. Figure 3.4 shows a portion of a scale-free network.

Recently, there has been work done using multiplex networks (e.g., see Matamalas et al. [2005] and Gómez-Gardenes et al. [2012]). The idea behind these networks is context and social ties affect individual behavior. For example, one may cooperate with family members but defect with co-workers. It is therefore necessary to have a representation that can adapt behavior for different social scenarios. A multiplex network has several layers. Each layer allows for different interactions and different strategy choices. Payoffs are then summed across all the layers. An example of a multiplex network is shown in Figure 3.5.

The claim is layering in a multiplex network provides a more holistic representation of an individual's behavior allowing for more sophisticated conduct in social dilemmas. The question then is, do real-world social dilemmas present a variety of different contexts? There is no empirical evidence indicating humans in a social dilemma experience various degrees of temptation

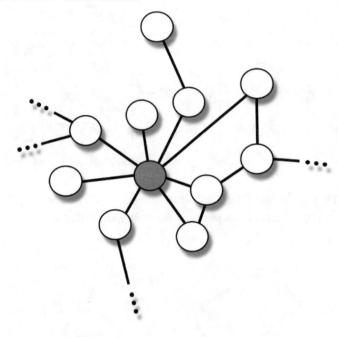

Figure 3.4: A portion of a scale-free network. The red vertex is a hub.

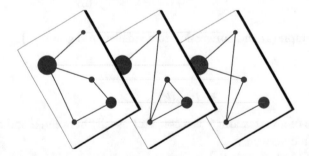

Figure 3.5: A multiplex networks with three layers. Red nodes indicate player cooperates in that layer. Node size is proportional to payoffs. Figure originally appeared in Matamalas et al. [2005]. Used with permission.

to defect (as suggested by Gómez-Gardenes et al. [2012]) and make their decisions accordingly. Players make decisions to cooperate or defect based solely on what choices other players make—not on the context. Indeed, a rational player is unlikely to continue cooperating with a defector regardless of whether that defector is a family member, a friend, or a complete stranger. Network performance depends on how strategies are updated. Allen and Hoyle [2017] discovered the frequency of cooperation in PGGs depends on whether strategies are synchronously or asynchronously updated. At this point it is unclear how (or even if) multiplex networks can lead to proximate explanations for human behavior in actual social dilemmas and the authors provide no guidance.

Remark 3.1   Spatial games are equivalent to network games on regular graphs. This can be seen by comparing Figures 3.1 and 3.2. If the edges are removed from the 2D grid, and players are required to interact with nearest neighbors in terms of Euclidean distance, the Von Neumann or Moore neighborhoods are identical. Either game can be converted into an instance of the other in $\mathcal{O}(N)$ time. Consequently, any statements made with respect to spatial games apply equally to network games and visa versa.

3.3   PROBLEMS WITH SPATIAL AND NETWORK GAMES

Let $\mathcal{N}(x)$ denote the neighborhood of a player x. By definition $x \in \mathcal{N}(x)$—i.e., x is a member of its own neighborhood. Also, if $y \in \mathcal{N}(x)$ then $x \in \mathcal{N}(y)$. Spatial games naturally make players members of multiple neighborhoods. Consider the lattice shown in Figure 3.6. The focal player x and it's four yellow neighbors form a Von Neumann neighborhood. The east neighbor of x is partially red because it is also in the Von Neumann neighborhood of a focal player z who has four red neighbors. Actually, every player in this lattice is a member of five dissimilar groups with Von Neumann neighborhoods or nine dissimilar groups Moore neighborhoods. (Every $i \in \mathcal{N}(j)$ is also a focal player of its own neighborhood with $j \in \mathcal{N}(i)$.) These groups are all of equal size and never change membership.

Grids are regular graphs. Santos and Pacheco [2005] mention regular graphs don't accurately represent human networking because regular graphs tend to lack long-range interconnections. They referred to a paper by Albert and Barabási [2002] claiming it described more appropriate real-world networks to investigate—although none of the interaction networks described involved social dilemmas. But, the real reasons why grid and spatial games are poor choices for studying cooperation actually has nothing to do with a lack of long-range connections.

The first problem with spatial games is players belong to multiple neighborhoods, which is something humans don't normally do. Coon [1946] observed that in primitive societies only family groups persist. Baumeister and Leary [1995] concluded people tend to devote more time and effort on fostering a limited number of close relationships rather than cultivating many more casual relationships. A computer model is supposed to help gain insight into the genesis of cooperation. That objective is made just a little bit harder the more the model deviates from human behavior. Humans do not deliberately form associations like those found in a lattice.

But there is a more serious problem with spatial games and it deals with strategy updates. Assume in each round of a spatial game a player engages in a PGG with his neighbors. Only

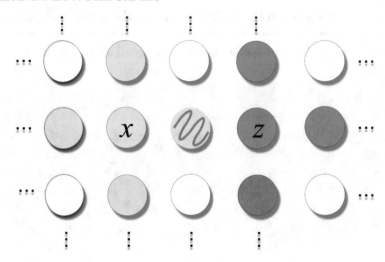

Figure 3.6: Two adjacent Von Neumann neighborhoods in a spatial game.

the focal player receives a payoff, not the entire neighborhood. That is, the focal player x receives the payoff from the PGG and not $\mathcal{N}(x)$. (Each neighbor of x is a focal player of its own neighborhood so each round every player in the game will receive a payoff.)

Remark 3.2 The focal player in an IPD game accumulates the payoff each round by competing in a one-shot pairwise game with each neighbor and then summing the individual payoffs. Some authors also have the focal player play a one-shot game with himself. This is pointless. Why would you ever defect against yourself?

Suppose in some given round a player x is tempted to adopt a neighbor y's strategy. x will adopt the strategy of y with a probability proportional to their payoff difference. Virtually all spatial PGGs use some form of strategy update process along this line. This decision process would make perfect sense if payoffs were accumulated only from a PGG among the focal player and his neighbors. Therein lies the problem.

Figure 3.7 shows two adjacent Von Neumann neighborhoods. Notice the payoffs players x and y receive each round are only partially determined by their pairwise interaction with each other. In fact, most of the payoff y receives comes from interaction with his blue neighbors—not the members of $\mathcal{N}(x)$. These blue players don't play a PGG with x. Thus, strategy updates based on payoff differences are not due to a superior strategy within a single PGG. Put another way, the payoff difference between x and y comes from playing two separate PGGs in which x and y

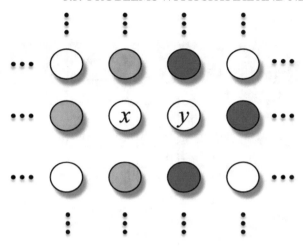

Figure 3.7: Two adjacent Von Neumann neighborhoods in a spatial game. The neighborhood of x consists of y and the red nodes and the neighborhood of y is x and the blue nodes.

are the only common players. The other players in these two games never interact. This situation is exacerbated with Moore neighborhoods.

The most serious problem with spatial games is virtually all of them that appear in the literature make predictions about long-term behavior that cannot be validated. As stated by Helbing and Yu [2010],

> ... the future of social experimenting lies in the combination of computational and experimental approaches, where computer simulations optimize the experimental setting and experiments are used to verify, falsify, or improve the underlying model assumptions.

Completely agree. Unfortunately, most research using computer models to predict cooperation levels does neither. Computer simulations are seldom validated with human experiments and results from human experiments are seldom used to modify and reevaluate the computer models.

Table 3.1 shows a sampling of spatial game models. Many of these models evolved cooperation levels in the population using a Monte Carlo process. (The issue of using Monte Carlo simulations is discussed at depth in a later chapter.) The model is run for a very large number of rounds to let things "settle down" and afterward actual data is recorded. Notice this transient period lasts tens of thousands of rounds in more than one model. Humans probably would not use a Monte Carlo process to modify their strategies. It is worth noting none of these authors described what process humans would use.

Equally problematic are the population sizes. A literature survey shows 100×100 lattice sizes (total population size 10,000) are quite common, but some models use population sizes an order of magnitude larger. How does one verify a model with that big of a population?

Table 3.1: A sampling of spatial games

Source	Lattice Size	Transient Rounds	Rounds/Game
Perc and Szolnoki [2008]	300 × 300	10^5	10^4
Hauert and Doebeli [2004]	100 × 100	5,000	10^3
Jin et al. [2018]	200 × 200	2.95×10^5	5×10^3
Rong et al. [2013]	100 × 100	5×10^4	5×10^4
Tanimoto [2015]	100 × 100	10^4	250
Szolnoki et al. [2009]	1,000 × 1,000	-	10^6
Chen et al. [2016]	200 × 200	4.5×10^4	5×10^3
Stivala et al. [2016]	100 × 100	-	10^9
Wu et al. [2018]	100 × 100	5×10^4	10^4
Menon et al. [2018]	128 × 128	-	10^4

An example helps illustrate the magnitude of this verification problem. The Szabó and Toke [1998] model used a 400 x 400 lattice. Orlando, Florida is the home of Walt Disney World. To verify their results, one would have to convince over half the population of Orlando to arrange themselves in a square lattice (with periodic boundary conditions), and then play thousands of pairwise PD games with themselves and their neighbors. The probability of that happening is zero. How useful are simulation results if they cannot be validated? Unvalidated results don't contribute anything to our understanding about cooperation.

The situation isn't much better with scale-free or small-world networks. For example, Wang et al. [2018] studied a PGG with a population size of 1,000 in a scale-free network. The simulation results were averaged over 1,000 rounds with 19,000 transient rounds. Cameron and Arias [2013] studied an IPD in small-world networks with the same population size and run from 70,000 up to 10^6 rounds. Neither of the models can be validated with a human population. Other graphs have been proposed but these border on the absurd. For instance, Taylor et al. [2007] studied bi-transitive graphs of various degrees. These included ring graphs and fully connected graphs but also included island models with fixed deme sizes and graphs with pentagonal cycles. It is hard to imagine a human population organizing themselves like that.

3.4   SUMMARY

Spatial and grid games are functionally equivalent, which means any advantages or disadvantages for one apply equally to the other. These two type of games don't contribute much (if anything) to our understanding of how cooperation levels change in human populations. They suffer from three primary problems.

- *The Von Neumann and Moore neighborhoods give false information about payoffs.*

  Most simulations update a player's strategy by comparing payoffs with players in their neighborhood. But the payoff of a neighbor $y \in \mathcal{N}(x)$ is obtained mostly via interactions with players outside of $\mathcal{N}(x)$.

- *The Von Neumann and Moore neighborhoods have artificial structures unlikely to ever be found in human populations.*

  It is difficult to envision humans forming associations like those in a Von Neumann or Moore neighborhood. Periodic boundary conditions would be nearly impossible to create and sustain unless the lattice is very small.

- *Spatial and grid game cannot be validated.*

  The games use Monte Carlo methods to evolve strategy frequencies. Humans would not do that. It is an open question how a different evolutionary process might affect simulation results. Population sizes of 10^4 and larger cannot be duplicated in human experiments. Humans will not participate in games that run for hundreds or thousands of rounds. Consequently, the spatial and grid games cannot influence experiment formats and experimental outcomes cannot improve the models.

CHAPTER 4

The Case Against Weak Selection

Social dilemmas are inherently competitive. Individuals interact repeatedly and acquire rewards. These rewards may benefit others or just the individual. Games by definition involve competition. It therefore understandable why social dilemmas can be formulated as games to study how and why cooperation levels vary in human populations.

Strategies in social dilemma games determine how a player reacts to the choices made by others. These strategies may change every now and then and do so through some evolutionary process. In other words, cooperation levels in the population evolve over time depending on the strategies in use. This evolutionary process follows—albeit, loosely—the neo-Darwinistic theory of survival of the fittest. Highly fit individuals survive. Survival, however, has a different interpretation in social games. In nature fitness measures the likelihood an individual will live long enough to reproduce. In social games games survival means a strategy will persist in the population (at least until the next round of the game). Players receive rewards as they play the game. These rewards may be monetary, an increase in social status or in some other form. Thus, survival in nature is genetic whereas in social dilemma games it is cultural. Nevertheless, fitness measures survival in both domains.

Every evolutionary process relies upon some form of selection process. Essentially, this selection process looks at the fitness level of each individual and decides who survives and who goes extinct. (With respect to social dilemmas survival and extinction refers to whether a given strategy respectively grows or decreases in the population over time.) Neutral selection assumes everyone in the population is equally fit so individuals are picked with equal probability. Most social dilemma games make the survival probability proportional to fitness. Two popular methods are frequency-dependent selection and roulette wheel selection.

Frequency (or abundance) refers to the fraction of a population using a given strategy. For example, if there are 4 cooperators and 6 defectors in a population of 10 players, then the cooperation frequency is 0.4 and the defection frequency is 0.6. Under frequency-dependent fitness an individual's expected fitness depends upon the population's composition—i.e., the frequency of a strategy determines the associated fitness of a player using that strategy. Positive frequency-dependent selection means the fitness increases as the frequency increases. Negative frequency-dependent selection has the opposite effect. Let x_i denote the frequency of strategy i and k the number of distinct strategies. Then the expected fitness of any player using strategy i

is $f_i = \sum_{j=1}^{k} x_j a_{ij}$ where $\sum_j x_j = 1$ and a_{ij} is the payoff whenever strategy i competes against strategy j. To show frequency-dependent selection suppose an N-player game has two strategies A and B. Then the probability of selecting a player with strategy A is given by

$$p_A = \frac{x_A f_A}{x_A f_A + x_B f_B}.$$

Roulette wheel selection makes the selection probability directly proportional to the total population fitness. That is, the probability of selecting an individual playing strategy i is

$$p_i = \frac{f_i}{\sum_{j=1}^{k} f_j}. \tag{4.1}$$

The name roulette wheel selection comes from making individual selection analogous to spinning a roulette wheel in a casino.

Remark 4.1   The roulette wheel concept is simple enough to describe by example. A game has 4 players indexed by $i = \{1, 2, 3, 4\}$ with fitness (payoff) values $f_i = \{10, 30, 20, 60\}$. The total population fitness is $f_{\text{tot}} = \sum_i f_i = 120$. Then the probability of selecting player i is $p(i) = f_i / f_{\text{tot}}$. The roulette wheel is constructed with 4 slots, but instead of each slot with the same width, as in roulette wheels found in casinos, the width is proportional to $p(i)$. For instance, the player 2 fitness is 25% of the total fitness. Therefore, the player 2 slot would be 90° wide. Player 4 has 50% of the total fitness so his slot would be occupy 50% of the wheel area. Then "spinning" the wheel chooses one of the players.

In practice, a roulette wheel is constructed by assigning each player i to a unique segment on the unit interval where the i-th segment length equals $p(i)$. A player is then selected by picking a random number between 0 and 1.

In some games fitness equals payoff, i.e., $f_i = \pi_i$. This won't work for roulette wheel selection—or any other method that translates payoff into a selection probability—if payoffs are negative. Fortunately, that situation can be easily handled by adding a positive offset to all of the payoffs to ensure $f_i > 0$ before constructing the roulette wheel. Another way to handle negative payoffs is with a linear mapping

$$f_i = 1 + \beta \pi_i, \tag{4.2}$$

where $\beta > 0$ is bounded to keep $f_i > 0$. Since either method can always be used, without loss in generality we will assume all payoffs are positive from now on.

Fitness can mean different things depending on the context. In nature fitness often results in an increased lifespan of an individual or greater reproductive opportunities whereas in evolutionary games it gives greater payoffs or increased strategy frequency in the population. Random drift and selection are two opposing forces in any evolutionary process. These forces are controlled in many models with a selection intensity parameter w. This parameter affects how much payoff contributes to an individual's fitness. Weak selection is $w \ll 1$.

Equating payoffs to fitness is regularly done, although many researchers map payoffs into fitness values. A linear mapping function is

$$f = 1 + w\pi, \tag{4.3}$$

where $w > 0$ is the selection intensity and π is the payoff. Equations (4.2) and (4.3) appear the same but w and β have completely different roles. In Eq. (4.2), β is chosen specifically to make f positive when payoffs are negative. For example, if $\pi = -2$ then $0 < \beta < 0.5$ keeps f positive. On the other hand, π is already assumed to be positive in Eq. (4.3) so w determines how much it affects fitness. The two parameters also have different bounds: $\beta \in (0, 1/|\pi|)$ while theoretically $w \in (0, \infty)$.

A more widely used linear mapping form is

$$f = 1 - w + w\pi, \tag{4.4}$$

where now $w \in [0, 1]$. Notice $w = 0$ makes the fitness equal to one which is a background fitness making selection neutral whereas fitness and payoff are equal when $w = 1$. Intermediate w values control how much payoff affect fitness.

Selection intensity can also be used to amplify payoff differences. Consider two players A and B with payoffs π_A and π_B, respectively. Let $\triangle\pi = \pi_A - \pi_B$. Then the probability p that strategy of player A replaces the strategy of player B is determined by the Fermi function from statistical physics

$$p = \frac{1}{1 + e^{-w\triangle\pi}}. \tag{4.5}$$

Large w represents strong selection because large payoff differences increase the probability player B changes strategy. On the other hand, weak selection has just the opposite effect since $w \to 0 \Rightarrow p \to 1/2$ (neutral selection) even with large $\triangle\pi$.

Weak selection was introduced by Nowak et al. [2004] and is widely used even today. That does not mean it isn't controversial. Indeed, weak selection has two major problems, one philosophical and one analytical.

Consider the IPD payoff matrix

$$\begin{array}{cc} & \begin{array}{cc} C & D \end{array} \\ \begin{array}{c} \text{payoff to } C \\ \text{payoff to } D \end{array} & \begin{pmatrix} b - c & -c \\ b & 0 \end{pmatrix}. \end{array} \tag{4.6}$$

Let $b = 3$ and $c = 1$. Then the payoff matrix for the row player is

$$
\begin{array}{c}
\phantom{\text{payoff to }C} \quad C \quad\ \ D \\
\begin{array}{c}
\text{payoff to } C \\
\text{payoff to } D
\end{array}
\begin{pmatrix}
2 & -1 \\
3 & 0
\end{pmatrix}.
\end{array}
\tag{4.7}
$$

There is a clear distinction between the cooperation and defection payoffs. Under weak selection Eq. (4.6) becomes

$$
\begin{array}{c}
\phantom{\text{payoff to }C}\quad\quad C \quad\quad\quad\quad\quad D \\
\begin{array}{c}
\text{payoff to } C \\
\text{payoff to } D
\end{array}
\begin{pmatrix}
1 - w + w(b - c) & 1 - w - wc \\
1 - w + wb & 0
\end{pmatrix}.
\end{array}
\tag{4.8}
$$

After substituting $b = 3, c = 1,$ and $w = 0.001$

$$
\begin{array}{c}
\phantom{\text{payoff to }C}\quad\ \ C \quad\ \ D \\
\begin{array}{c}
\text{payoff to } C \\
\text{payoff to } D
\end{array}
\begin{pmatrix}
1.001 & 0.998 \\
1.002 & 0
\end{pmatrix}.
\end{array}
\tag{4.9}
$$

These values are sufficient to satisfy the $T > R > P > S$ criteria discussed in Section 2.2. However, now the distinction between cooperating and defecting is not so evident. Effectively weak selection eliminates the social dilemma entirely because the payoff to the row player depends mostly on the opponent's choice. It therefore doesn't matter whether the row player chooses C or D which amounts to neutral selection. Indeed, one could argue that with such a payoff matrix any rational player should always cooperate—which directly contradicts experience with any real-world social dilemma.

From a philosophical perspective weak selection makes no sense. People play games with the intention of winning. This is true whether the game is chess, baseball, or a social dilemma game. Players pick game strategies that hopefully provide them some advantage over any opponent(s). (The same is true in real-world social dilemmas.) The problem is weak selection in the limit essentially produces neutral selection because all strategies have roughly the same fitness. Neutral selection blurs any distinction between different strategies; higher payoffs provide little or no advantage. For example, by Eq. (4.4) with $w = 0.001$, a payoff $\pi_A = 20$ has a fitness of $f_A = 1.019$ while a payoff $\pi_B = 30$ (50% higher) has a fitness of $f_B = 1.029$. Roulette wheel selection would have nearly the same width slots for these two strategies. What is the point of playing a game if it doesn't matter what strategy is used? Yet this is precisely what happens under weak selection. As $w \to 0$ selection becomes neutral. Put another way, if weak selection reduces or eliminates any gains, what is the point of cooperating? Yet weak selection is ubiquitous in computer models. Why?

The main reason given is weak selection linearizes the mathematics making analysis easier. For example, Wild and Traulsen [2007] gave a formula for computing the fixation probability of a single mutant in a population. They derived an approximate form to make the analysis simpler by assuming weak selection and then doing a Taylor series expansion. Sample and Allen [2017] state that weak selection, in conjunction with large population sizes, allows many results to be expressible in a closed form that would not be otherwise possible.

But mathematical convenience is not a compelling reason for using weak selection— especially if it alters the game dynamics. There is a lot of misunderstanding and confusion surrounding the use of weak selection. There is also no consensus on its usefulness.

Ohtsuki et al. [2007] give a non-mathematical justification for weak selection:

> "... in most real life situations we are involved in many different games, and each particular game only makes a small contribution to our overall performance."

While on the surface that statement may be true, the study of cooperation in a social dilemma has a very restricted context. That is, individuals are engaged in only one "game" and their behavior in this one particular game is the only thing of interest. Strategies among players determine an individual's fitness in the social dilemma. The whole purpose of studying the social dilemma is to identify conditions under which specific strategies prevail. It is therefore irrelevant how an individual behaves in some other environment or what factors in these other environments affect overall behavior.

An examination of some claims made in a paper by Fu et al. [2009] will help illustrate where this confusion is coming from. One claim is weak selection can be justified

> "[because] the weak selection limit has a long tradition in theoretical biology ... when working with population genetics."

Comparing population genetics with social dilemmas is like comparing apples and oranges. Both involve populations and natural selection plays a role in each. But that is where the similarity ends.

Population geneticists study gene frequencies (genetic evolution) while game theorists study strategy frequencies (cultural evolution). Gene allele frequencies vary through various physical processes such as mating and random mutation. Strategy frequencies vary through completely different mechanisms and identifying them is key to understanding the genesis of cooperation in humans. Individuals may choose a different strategy after performing a cost/benefit trade-off analysis, which suggests some underlying cognitive process. Studies indicate emotional responses such as envy [Parks et al., 2002], guilt [Greenwood, 2015a, Miettinen and Suetens, 2008], and anger [Castagnetti et al., 2018, Seip et al., 2014, Weber et al., 2018, Wubben et al., 2009] can stimulate cooperation. Clearly, genetic evolution and cultural evolution operate under different governing dynamics. As stated by Traulsen [2010],

> "Although weak selection seems to be relevant for genetic systems, when it comes to cultural evolution it may not be an appropriate approximation."

Another claim by Fu et al. [2009] is

"indeed, inclusive fitness analysis exclusively relies on the assumption of weak selection"

and for supporting evidence they cite the work by Taylor et al. [2007]. This is not a convincing argument. The problem is Taylor et al. studied cooperation in transitive graphs such as graphs with interconnected pentagonal cycles and island structures with size-4 demes—structures which seem an unlikely organization for human populations.

"The results derived from weak selection often remain as valid approximations for larger selection strength"

is another Fu et al. [2009] claim. A growing body of empirical evidence indicates just the opposite is true. One of the best counter-arguments was provided by Wu et al. [2013]. They considered a public goods game where $N = 100$. Each round 5 players were randomly selected and given an opportunity to contribute $c = 1$ monetary unit to a common pool. Cooperators contribute. This pool was then multiplied by $r = 3$ and then redistributed to all players whether or not they contributed. Some players who contribute could also choose to punish defectors who did not contribute by imposing a fine $\alpha = 1$ on them but pay a cost $\gamma = 0.3$ to do so. Players must subtract and contributions or costs from their return. Numerous simulations were run for different weak selection values. (In their work β represents selection intensity.)

Figure 4.1 shows the ranking of the three strategy types: cooperators, punishers, or defectors. Ranking refers to which strategy dominates; the highest rank is for the strategy with the highest frequency. Panel A shows the results when payoffs were exponentially mapped, $f = \exp(\beta\pi)$ and Panel B when the payoffs were linearly mapped, $f = 1 - \beta + \beta\pi$. The dashed lines are weak selection approximations. Punishers dominate under weak selection but defectors dominate at higher intensity levels. The vertical dashed lines show selection intensity values where the rankings changed. These results clearly show weak selection results do not hold at even moderate selection intensity values.

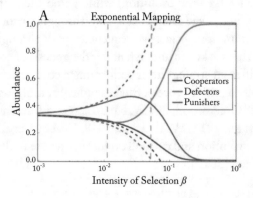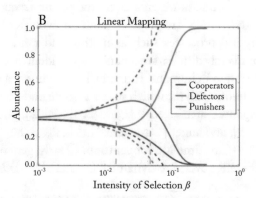

Figure 4.1: Abundance (frequency) of the three strategies as of function of selection intensity (β). (See text.) This figure originally appeared in Wu et al. [2013]. Used with permission.

Researchers are split on whether weak selection results hold at higher selection intensities. Mixed messages lead to confusion about weak selection. Some researchers (e.g, Ohtsuki et al. [2007]) assert results derived from weak selection are valid at higher selection rates. For instance, Fu et al. [2009] said

"... the results derived from weak selection often remain as valid for larger selection strength."

Others believe just the opposite is true. Langer et al. [2008] published a paper on spatial invasion where they observed

"... under weak selection random drift dominates, which makes it much harder to extract characteristic features of the evolutionary process. In contrast, our approach based on strong selection facilitates clear-cut conclusions."

Clearly, the Fu et al. [2009] and Langer et al. [2008] papers have opposite viewpoints on whether weak selection results extend to higher selection intensity levels. Ironically, two authors on the first paper are authors on the second paper as well! You can't have it both ways. Either weak selection results are broadly applicable or they are not.

Wu et al. [2010] point out

"... under the assumption of weak selection, some important insights can be obtained analytically. It has to be pointed out that these results do, in general, not carry over to stronger selection."

Even human experiments do not support using weak selection predictions. Traulsen et al. [2010] conducted experiments with university students playing PD rounds. They found the selection intensity level among humans was surprisingly orders of magnitude higher than weak selection levels. Consequently, this led them to conclude

"[selection intensity] β is also so high that analytical results obtained under weak selection may not always apply."

Juxtaposing the two papers by Wu et al. [2013] and Traulsen et al. [2010] should set off alarm bells among the proponents of weak selection. The Wu et al. paper provides convincing analytical evidence that weak selection predictions are invalid at higher selection intensities. The Traulsen et al. paper found humans use selection intensities orders of magnitude higher than weak selection. One could logically conclude any conclusion made under weak selection most likely doesn't apply in human groups. What then is the justification for still using weak selection analysis in social dilemma games? Mathematical convenience is not a good enough reason— particularly if it produces questionable results. The burden of proof is on the weak selection community to provide more credible arguments for its continual use. Until then, it should not be used in social dilemma research work because any conclusions derived from its use may be invalid.

4.1   SUMMARY

- Weak selection was introduced by Nowak et al. [2004] and has been extensively used since then.

- Analytic results show weak selection predictions often are invalid at higher selection intensity levels.

- Experiments show selection intensity levels are orders of magnitude higher in human populations.

- Weak selection should not be used in social dilemma games until its proponents provide convincing evidence of its applicability in real-world situations.

CHAPTER 5

The Moran Process and Replicator Dynamics

Image a population of $N > 2$ players participating in some economic activity. Each player has a strategy that determines how they will interact with other players. This strategy is invariant in the sense the same strategy is used regardless of who they interact with. From time to time a player may switch to a different strategy in the hope it will produce a higher payoff. Strategies therefore change over time. Intuitively, this implies some process of evolution is taking place. In social dilemma games this evolutionary process is usually either a *Moran process* or *replicator equations*. In this chapter those two processes are defined and compared.

5.1 DEPICTING STRATEGY EVOLUTION

Strategy evolution is depicted as trajectories in an n-simplex. Before giving some examples several geometry-related terms need to be defined. These definitions are not meant to be mathematically rigorous, but are sufficient for our purposes.

Definition 5.1 (convex set) *A set $S \subseteq \mathbb{R}^n$ is convex if for all $x, y \in S$ and $\lambda \in [0, 1]$, then $\lambda x + (1 - \lambda)y \in S$.*

Convex   Non-Convex

Definition 5.2 (convex hull) Given a set of points $S \subseteq \mathbb{R}^n$, the convex hull is the smallest convex set containing S.

Definition 5.3 (n-simplex) An n-simplex is a convex hull with $n + 1$ extremal points (vertices).

The simplex will always have "flat" sides. In a regular simplex all edges have the same length. Only regular simplices are used in games research. A 0-simplex is a point. Figure 5.1 shows a 1-simplex and a 2-simplex.

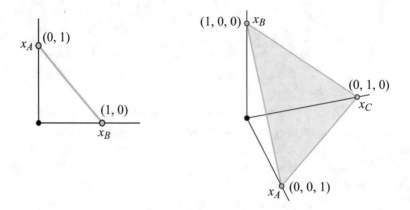

Figure 5.1: A 1-simplex (left) and 2-simplex (right). The simplexes are shown in green. These are $n - 1$ subspaces in an n-dimensional space.

Observe that $\sum_i x_i = 1$ holds at every point in the simplex. This characteristic is important because trajectories in a simplex track how strategy frequencies change over time (see Figure 5.2).

Remark 5.4   Let $\triangle^n$ denote an n-simplex. A degenerate mapping f is $f : \triangle^{n-1} \to \triangle^n$. The simplices of interest here are nondegenerate—i.e., $\triangle^n$ is not the image of any degeneracy map.

There is a relationship between trajectory shape and population size. Although every point in the simplex represents a valid set of strategy frequencies—i.e., at every simplex point $\sum_j x_j = 1$—not every point can necessarily lie on a trajectory. A strategy frequency is defined as $x_i = n_i/N$ where N is the population size and n_i is the number of strategy i players. Notice x_i is always a rational number. Consequently, all possible frequencies are not necessarily valid for a given population size. For example, the interior point in a 2-simplex $\{x_A, x_B, x_C\} = \{0.359, 0.221, 0.420\}$ represents a possible strategy frequency set, but only if N is sufficiently large. (In this case the strategy frequencies would be rational numbers only if N is an integer multiple of 1000.)

Trajectories can move only through those simplex points representing rational number frequency sets. Smooth trajectories, such as those shown in Figure 5.2, move through points of arbitrary precision which means they exist if and only if the population size is infinite. In practice,

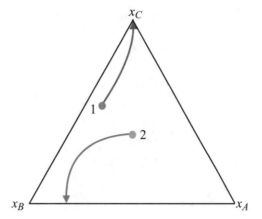

Figure 5.2: A 2-simplex showing two cases of strategy evolution. A, B, and C are pure strategies and x_i their corresponding frequency. Each point in the simplex corresponds to a unique set of strategy frequencies. The blue trajectory starts at point 2 and terminates on the x_A–x_B boundary. Thus, the population initially contained all three strategies but ends with only A and B strategies (mostly strategy B). The red trajectory starts at point 1 and terminates at the x_C vertex. This population ends up with the entire population using strategy C.

however, population sizes are finite which means only a subset of simplex points represent valid frequency sets. An example is shown in Figure 5.3. Trajectories are constrained to move only between the points shown. This means trajectories for finite populations are actually piecewise linear. Another difference is some interior simplex points may become trajectory fixed points in a finite population simplex. On the other hand, no interior points in the infinite population simplex are fixed points; any fixed points are at a vertex or on the simplex boundary.

5.2   THE FREQUENCY-DEPENDENT MORAN PROCESS

Consider an N-player game that has two strategies A and B with a symmetric payoff matrix

$$
\begin{array}{cc}
 & \begin{array}{cc} A & B \end{array} \\
\begin{array}{c} A \\ B \end{array} & \left(\begin{array}{cc} a & b \\ c & d \end{array} \right).
\end{array}
$$

Fitness equals accumulated payoffs. This fitness is frequency dependent. That is, if i players choose strategy A, and self-play is not allowed, then the expected fitness for a player using strategy A is

$$
f_i = \frac{a(i-1) + b(N-i)}{N-1},
$$

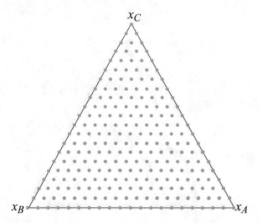

Figure 5.3: A 2-simplex for a population of size $N = 20$. The points shown are the only strategy mixtures where there are an integer number of strategies that total to 20. Trajectories can only move in a piecewise linear fashion between these points.

and for a player using strategy B

$$g_i = \frac{ci + d(N - 1 - i)}{N - 1}.$$

The Moran process is a stochastic evolutionary process where at most one strategy in the population can change per round. It was originally developed by Moran [1958] to describe the change in gene frequencies in a population. In social dilemma games it is used to describe how strategy frequencies evolve in a finite population. No mutations are allowed. Since there is no mutation strategies cannot spontaneously appear in the population. If a strategy disappears from the population it cannot later reappear.

The strategy change mechanism is referred to as an *update rule*. In a well-mixed population the update rule is simple. Choose two individuals randomly, the first proportional to fitness and the second with equal probability. The first individual reproduces and its offspring replaces the second individual. This keeps the population size constant. It is straightforward to calculate the probability of getting an A strategy increase in a given round. This is just the probability of picking the 1st player who uses strategy A times the probability of picking a second player who uses strategy B. If i out of the N players are using A, then

$$\text{Prob}(A \text{ increase}) = \frac{if_i}{if_i + (N - i)g_i} \times \frac{N - i}{N}. \tag{5.1}$$

Four update rules are commonly used in spatial games.

1. *Birth-Death* (BD)—A player is randomly picked out of the population proportional to fitness and reproduces. The offspring replaces a randomly chosen neighbor.

2. *Death-Birth* (DB)—A player is randomly picked out of the population to die. The neighbors then compete to reproduce and the offspring replaces the player that died.

3. *Imitation* (IM)—A player is randomly chosen for a strategy update. A competition between the chosen player and his neighbors determines the replacement strategy.

4. *Pairwise-Comparison* (PC)—A player and one neighbor are randomly picked. The focal player adopts the neighbor's strategy with a probability p tied to fitness differences. The Fermi function from statistical physics is used. A focal player using strategy A adopts strategy B from a neighbor with probability

$$p = \frac{1}{1 + e^{-w(\delta_f)}},\qquad(5.2)$$

where $\delta_f = f_B - f_A$ and w is the selection intensity.

There is a possibility no strategy change takes place whenever a PC or an IM update rule is used. The competition used in the update rules could be deterministic, by choosing the highest fit individual, or stochastic using roulette-wheel selection. A Moran process does require only positive fitness values which means all payoffs should be positive. But, negative payoffs are not a problem because they can be rescaled as was shown in Eq. (4.2).

5.3 REPLICATOR EQUATIONS

Suppose there are m strategies used by a population of $N > 3$ players. Let n_i be the number of players using strategy i at time t. The number of players using i evolves according to $\dot{n}_i = n_i \pi_i(x)$ where $\dot{n}_i$ is the derivative dn_i/dt, $\pi_i(x)$ is the payoff for using i and x is the population state. If $x_i = n_i/N$ is the frequency of strategy i, the population state is $x = [x_1, \ldots, x_m]$. With $N = \sum_{j=1}^m n_j$ being the total population size, the rate of change in the frequency of strategy i is

$$
\begin{aligned}
\dot{x}_i &= \frac{d(n_i/N)}{dt} \\
&= \frac{N\dot{n}_i - n_i \dot{N}}{N^2} \\
&= \frac{N\dot{n}_i - n_i \sum_{j=1}^m \dot{n}_j}{N^2} \\
&= \frac{\dot{n}_i}{N} - \frac{n_i}{N} \sum_{j=1}^m \frac{\dot{n}_j}{N} \\
&= \frac{n_i \pi_i(x)}{N} - \frac{n_i}{N} \sum_{j=1}^m \frac{n_j \pi_j(x)}{N} \\
&= x_i \pi_i(x) - x_i \sum_{j=1}^m x_j \pi_j(x).
\end{aligned}
$$

The summation $\sum_{j=1}^{m} x_j \pi_j(x)$ is the average payoff in the population, which for convenience is denoted by $\bar{\pi}$. Thus, the replicator equations are a set of m first-order differential equations of the form

$$\dot{x}_i = x_i \left[\pi_i(x) - \bar{\pi} \right]. \tag{5.3}$$

Replicator equations indicate how strategy frequencies change over time. They express the principle of natural selection mathematically. A strategy frequency increases if the RHS of Eq. (5.3) is positive, decreases if negative, and does not change if zero. In other words, if the payoff for using strategy i is greater than the average population payoff then strategy i increases in frequency. It decreases if the payoff is less than the average and does not change if equal to the average. It is worth noting replicator equations only indicate aggregate population behavior and do not provide any information about an individual player's behavior.

Remark 5.5　$\dot{N} = 0$ because the population size is fixed. By definition $\dot{N} = \sum_{j=1}^{m} \dot{n}_j$. The summation equals zero if $\dot{n}_j = 0 \; \forall j$. However, if some $\dot{n}_k$ increases ($\dot{n}_k > 0$) then other $\dot{n}_j$ terms must decrease ($\dot{n}_j < 0$) to keep the summation equal to zero. This occurs naturally because the set of first-order differential equations are coupled through the average payoff term $\bar{\pi}$.

The replicator equation functions must be everywhere differentiable which implies an infinite population size. Unfortunately, there is no such thing as an infinite population in nature. This has given rise to the discrete replicator equations, which are a set of coupled first-order difference equations. As before, fitness equals accumulated payoffs.

Let k_i, $i = 1, 2, \ldots, m$ represent the number of players using strategy i in a finite population of size N. Then $x_i^t = k_i/N$ is the frequency of strategy i at time t. The discrete replicator equations are of the form

$$x_i^{t+1} = x_i^t \left(\frac{\pi_i^t}{\bar{\pi}^t} \right), \tag{5.4}$$

or, equivalently after multiplying both sides by N

$$k_i^{t+1} = k_i^t \left(\frac{\pi_i^t}{\bar{\pi}^t} \right), \tag{5.5}$$

where π_i^t is the payoff for strategy i at time t and $\bar{\pi}^t$ is the average population payoff. The term in parenthesis is the ratio of a strategy's fitness to the average population fitness. k_i increases in the next round if this term is greater than one; decreases if less than one; and remains unchanged if equal to one.

Clearly, $\sum_i k_i = N$ is required. The problem is there is no guarantee the term in parenthesis is an integer which means the LHS of Eq. (5.5) may not be an integer. To overcome this problem the quantization algorithm below is used. This algorithm, first introduced to social dilemma games by Greenwood et al. [2016], takes x_i^{t+1} and N as inputs and returns k_i' where $\sum k_i' = N$.

Algorithm Quantization

1: Let q be the number of strategies.
2: Compute

$$k_i' = \left\lfloor N x_i + \tfrac{1}{2} \right\rfloor \quad , \quad N' = \sum_i k_i'.$$

3: Let $d = N' - N$. If $d = 0$, go to step 5. Otherwise, compute the errors $\delta_i = k_i' - N x_i$.
4: If $d > 0$, decrement the d k_i'''s with the largest δ_i values. If $d < 0$, increment the $|d|$ k_i'''s with the smallest δ_i values.
5: Return $\begin{bmatrix} k_1' & k_2' & \dots k_q' \end{bmatrix}$ and exit.

No further strategy evolution takes place if the trajectory hits a fixed point. In both the finite and infinite population replicators simplex vertices are fixed points because there is some $x_i = 1.0$, all other $x_j = 0$ and no mutation is allowed. At interior points every $x_i > 0$. This means there are no interior fixed points except at points where all payoffs are equal. Unfortunately, in the finite population model it is possible to have artificial fixed points due to quantization errors. These artificial fixed points arise whenever the payoff-to-average payoff ratio on the RHS of Eq. (5.5) does not equal 1.0, but it is not large enough or small enough to change k_i^{t+1}.

> **Remark 5.6** The finite population trajectory can escape an artificial fixed point by incrementing strategies with a positive RHS and decrementing those with a negative RHS (to keep the population size constant) post quantization. Of course this additional step is unnecessary if the trajectory is not at a fixed point or if $\pi_i / \bar{\pi} = 1 \; \forall i$.

5.4 COMPARING EVOLUTIONARY MECHANISMS

In the early 1980s, the personal computer became readily available. Research scientists and engineers could now conveniently use the powerful Monte Carlo techniques, developed decades earlier, to conduct stochastic searches for good solutions to difficult physical chemistry and particle physics problems—without leaving their desk. The predominate Monte Carlo search technique at the time was simulated annealing (SA). The obvious and immediate appeal for SA was threefold: it was broadly applicable; it was easy to explain; and it was trivial to program.

Remark 5.7   Within a few years SA was surpassed by the evolutionary algorithms (genetic algorithms, differential evolution, etc.) which produced superior stochastic search results with less computational effort.

The Moran process was originally developed to describe gene frequency evolution. It was introduce to social dilemma games by Nowak et al. [2004] and quickly achieved widespread use for the same reasons SA did in the 1980s: it could be used in any social dilemma game; it was easy to explain; and trivial to program. Even game theorists without sophisticated programming skills could quickly and easily get a Moran process coded and running. But easy in coding is not a sufficient justification for its widespread use. Indeed, the Moran process suffers from two fundamental problems.

The first problem is the Moran process is a purely stochastic process. Strategy changes require randomly choosing two strategies. Unfortunately, random sampling does not guarantee repeated outcomes and this becomes more problematic with smaller populations. To show just one example of the problem associated with random sampling a 40-player PGG with BD updating was constructed. Cooperator, defector, and punisher were the three strategy choices. Figure 5.4 shows two runs that predicted completely different outcomes even though they started with the same initial strategy frequencies. One run predicted $x_D = 1$ (everybody defects) while the other run predicted $x_D = 0$ (nobody defects). Only one thing was different in the two runs: they used different seeds for the random number generator.

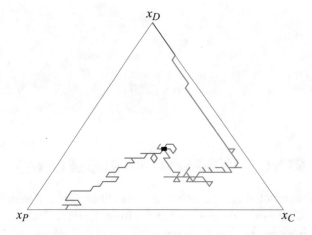

Figure 5.4: Two runs of a 40-player PGG using BD updating. The initial frequency distribution was $x_C = 0.35$, $x_D = 0.325$, and $x_P = 0.325$ in both runs. The start of the two runs is indicated by the black rectangle.

The second fundamental problem with the Moran process is even more severe: it doesn't match the way people behave in competitions. Consider a group of players who are competing against each other in some game. If they make decisions using a Moran process, then each player knows the following rules apply.

At some point during the game you may be forced to change your strategy. You will not be told ahead of time when or even if this will occur. Your approval is irrelevant. It is not needed because the strategy change is mandatory. Moreover, you will have no say in what your new strategy is. You will be told what new strategy to use.

Who plays a game—any game—under those terms? The game could be baseball, GO, or some video game. Players want to make their own decisions about what strategy to use. This desire is especially true if financial gains and losses are involved—which covers just about every real world social dilemma. Consider the funding for National Public Radio (NPR) who relies heavily on listener contributions. Cooperators contribute while defectors do not. Yet anyone can listen, including defectors. But under a Moran process all decisions are made without player consent. If the Moran process were used to pick cooperators and defectors then an individual could be forced to contribute money to NPR whether they wanted to or not. This would most certainly end up in court! People don't make decisions in situations involving finances by what amounts to coin flipping. Consequently, the Moran process can't provide any intuitive understanding about human cooperation.

On the other hand, replicator dynamics are strictly deterministic. Replicator equations can provide proximate answers to why cooperation grows or dies out because, at least in one respect, they do mimic human behavior. Under replicator dynamics above average payoff strategies will increase in the population and below average payoff strategies will decrease. Players may not know exactly what strategy an opponent may be using but they can tell whether or not it is effective by observing the payoffs. They may even be able to infer what strategy is being used against them. Regardless, there is always a temptation to switch to a better performing strategy. Replicators do mimic such behavior because strategy changes depend solely on how well a strategy's fitness compares to the average population fitness.

People study social dilemma games to help identify conditions that promote cooperation in human populations, particularly among non-kin. Any insight obtained from these games is meaningful—and, therefore useful—only if the game's governing dynamics correspond to actual human behavior or decision making. The Moran process fails to do so. All strategy changes are the outcomes of a random process and all changes are mandated without player consent. Well, people don't randomly make choices in social dilemmas because they almost always have financial consequences. Nor are they willing to let others control what they can and can't do in some competition. Consequently, it seems highly unlikely the continued use of a Moran process to control strategy changes will ever help reveal the proximate causes of human cooperation. It is therefore recommended that researchers stop using a Moran process in social dilemma games.

Researchers need to use methods that more closely emulate human motivations, behaviors, or decision-making processes.

5.5 SUMMARY

- Both the Moran process and replicator equations indicate only aggregate population behavior and provide no information about individual player behavior.

- The Moran process evolves strategies via a stochastic process. Replicator equations evolve strategies deterministically.

- The differential equation form of replicator equations assumes an infinite population size; the discrete form assumes a finite population size.

- The Moran process is incapable of modeling human behavior and therefore cannot provide insight into why humans cooperate. Consequently, its continued use in social dilemma games is not recommended.

CHAPTER 6

The Problems with Computer Models

Nature is a collection of physical systems. These systems are diverse and can be biological, electrical, chemical, mechanical, geological, or any combination thereof. We can observe and record a system's response to different environmental conditions and make conclusions about its intrinsic properties and characteristics. The objective is to know as much as possible about the system so that we can reliably predict how it will behave in specific circumstances that have not been observed.

In many cases it is difficult or implausible to create the inputs needed to see how a physical system will respond in a specific situation. Instead, we rely on computer models. If the computer model (sometimes called a mathematical model) accurately captures the system's properties, then we can use the model to make predictions about the system's behavior.

There are two broad classes of computer models: analytical and agent-based. Analytical models describe the system behavior with mathematical equations. These equations can then be solved to extract the system behavior. For example, electrical and mechanical systems can be described by nth order linear ordinary differential equations (continuous systems) or difference equations (discrete systems). Agent-based models contain individual entities that interact with other entities. Individual agent responses can then be aggregated to form conclusions about the overall system. Social dilemmas can be thought of as a physical system with cultural dynamics. They are studied using social dilemma games which are basically just a computer model. Both types of computer models are used. Replicator equations are analytical models while spatial games are agent-based models.

Social dilemma games are created to get an answer to the most important open question in nature: why do unrelated humans cooperate and under what conditions will this cooperation be sustained or even grow? Over the years literally thousands of papers have been published on this topic, many with an accompanying computer model. Yet, despite all of the research, all of the papers published, all of the computer time spent running simulations, one could credibly argue our understanding of cooperation and its origin is not much better today than it was 20 years ago. Why not? There are three primary reasons: (1) the models are not realistic, (2) the models are not validated, and (3) the models are used to answer the wrong questions. In this chapter each of these reasons will be examined.

6.1   MODEL REALISM

Social game models cannot help understand why cooperation is so pervasive if they contain un-realistic features that don't emulate human behavior. In a previous chapter unrealistic population sizes and the extraordinary number of game rounds was discussed (c.f., Table 3.1). But they are not the only unrealistic aspects of many social dilemma game models.

The first problem area requires another look at the Moran process. Here we focus on the pairwise comparison (PC) update rule, but similar arguments apply to the other update rules as well. Recall in PC updating two players are randomly chosen [Traulsen et al., 2006]. Suppose the first player uses strategy A while the second uses strategy B. Then the probability p the first player switches to strategy B is given by the Fermi function

$$p = \frac{1}{1 + e^{-w(\pi_B - \pi_A)}}, \tag{6.1}$$

where w is the selection intensity and π_A, π_B the respective payoffs. This update rule is illogical. If a player is thinking about choosing a different strategy, why look at only one other player? And why a randomly chosen player? A rational player would look at several players, ignore those doing poorly, and then decide what to do. Moreover, with finite w there is a non-zero probability of switching to strategy B *even if it does worse than strategy A*. Why would a rational player do that? No justification is given.

A second problem area is most game theorists conflate the player's choice, e.g., C or D, with the player's strategy. The justification is if a player switches from say C to D then there must have been a strategy change. Choices are not the same thing as strategies. A good example is tit-for-tat (TFT) which is the best known strategy for the 2-player iterated prisoner's dilemma game. Under TFT a player cooperates the first round and thereafter chooses whatever his opponent did in the previous round. Thus, a player using TFT could pick D in one round and C in another round. Two different choices, but only one strategy. Choices are the actions dictated by a strategy, not the strategy itself. The whole purpose of studying social dilemma games is to decipher strategies to identify why humans cooperate. This is not possible if the choice and the underlying strategy are considered the same thing.

Nowak [2006] posited five rules that promote cooperation. One of those was direct reciprocity where future choices are tied to past choices. For instance, an individual is far more likely to cooperate with someone who has cooperated with them in the past. People can relate to direct reciprocity. We tend to treat others the same way they treat us. One of the major problems with many social dilemma models is the absence of direct reciprocity. This makes the model unrealistic. An example will help fix ideas.

Consider a 100×100 spatial game with Moore neighborhoods. Every round focal players compete in a pairwise prisoner's dilemma game with their neighbors. Updates are done via a BD Moran process. Only one strategy update in the population occurs each round and, given the population size, it is reasonable to assume some focal player and his neighbors could play many rounds and none of them gets picked for a strategy update. The problem is the computer model is constructed so that a player must use the same choice against every player in his neighborhood. This leads to situations where, for example, a player continues to cooperate with his neighbors regardless of whether they cooperate or not with him. No rational player would continue to be exploited round after round by cooperating with an opponent who consistently defects. Humans do not behave that way. Yet virtually all spatial game computer models don't allow different choices with different players. Put another way, there is no provision for direct reciprocity built into the models. It is worth noting that this forced behavior with every opponent problem is also prevalent in well-mixed population computer models.

6.2   MODEL VALIDATION

Computer models should act just like the system they are designed to imitate. When they do, the model can accurately predict how the physical system will behave. Inaccurate models are worthless. Model validation is a process to ensuring model/system accuracy. This concept is expressed in the first principle of computer modeling:

> Computer model predictions shall not be presumed credible unless and until the model has been properly validated

Significant problems can happen by using an unvalidated model. Anthropogenic climate change is a topic of enormous interest (some would call it an obsession) because greenhouse gases accumulating in the atmosphere absorb heat from the Earth's surface. This heat then gets re-radiated back to the surface causing global warming. Predictions on the long-term effects on the ecosystem from climate change come exclusively from computer models. Model accuracy—which can only be determined after thorough validation testing—is therefore essential. Unfortunately, so far climate model validation has not gotten sufficient attention by climate scientists [Baumberger et al., 2017]. This is unacceptable. Trust in climate models is absolutely crucial because model predictions are influencing public policy decisions. Validation builds trust in a model. Using predictions from unvalidated models is likely to produce ineffective and costly public policies.

Algorithm 6.1 shows the steps involved in a proper model validation. After forming a hypothesis the model structure and parameter set are defined. In social dilemma games the structure would be a well-mixed population, a network, or a spatial game. The parameter set influences the model's behavior. The parameter set includes things like the population size, neighborhood size, player interactions, strategy update rules, payoff-to-fitness mapping, and so forth. The physical system environment is represented in the computer and then simulations are run.

Algorithm 6.1 Model validation

1: Formulate a hypothesis about a physical system
2: Design a model structure and parameter set to implement the hypothesis
3: Run simulations
4: Collect model results
5: Compare model results with observed results
6: If model results and observed results ϵ-close, then exit
7: Update parameter set and go to STEP 3

Observed data is recorded from human experiments or collected from real-world social dilemmas. The observed results and model results are then compared and the model is validated if they are in some sense ϵ-close. If not, the parameters are adjusted accordingly and more simulations are run. This process is repeated until the model and physical system results agree. At that time the model can be used to predict how the physical system will behave in specified environments.

Data from real-world social dilemmas is not easy to find. The good news is numerous social dilemma games with human subjects have been conducted. It is worthwhile to take a look at a few of them. Traulsen et al. [2010] arranged subjects into a virtual spatial grid with periodic boundary conditions and Von Neumann neighborhoods. Each group played 25 rounds of a PD game. C or D were the strategy choices. They found initially players use the imitate-the-best rule where they adopted the best performing strategy in the group. This tendency broke down over time. They did find an unanticipated result. A control experiment was also conducted in which players were reassigned each round. A spatial structure has fixed neighbors which gives players the opportunity to form cooperative clusters. Reassigning players each round prevents cluster formation. Surprisingly, there was no significant difference in the cooperation levels. They did find players imitate strategies of others with a probability that increases as the payoff difference increases.

In the Rand et al. [2014] study subjects were arranged in a ring and interacted with one, two or three neighbors ($k = 2, 4,$ or 6) on each side and played repeated rounds of a PGG. C and D were the only strategy choices. The experiment had two parts: in the first part the network was static while in the second part subject positions were shuffled prior to each round. The cost/benefit payoff matrix (Eq. (2.7)) was used. They found $b/c > k$ was a necessary condition for cooperation to succeed. More specifically, defectors with all defecting neighbors switched to cooperation 15.7% of the time when $b/c \leq k$, but 17.4% of the time when $b/c > k$. Conversely, a cooperator with all cooperating neighbors switched to defection 14.1% of the time when $b/c \leq k$ but only 5.1% of the time when $b/c > k$. This latter outcome is somewhat surprising; a larger defection rate seems likely. They speculated players tend to focus more on the payoffs others received rather than reciprocating neighbors' behavior. Also, the authors did point out only one cooperating neighbor is needed to break even when $b/c > k$ is satisfied which does provide a plausible explanation.

Cuesta et al. [2015] conducted 24 IPD experiments involving 243 subjects. Players were initially arranged in a ring network with links to nearest and next-nearest neighbors, but subsequently were allowed to choose up to five new opponents. C and D were the only strategy choices. The PD payoff matrix was "weak." (In the standard PD payoff matrix (Eq. (2.4)) $T > R > P > S$. The payoff matrix is weak if $P = S$.) In static networks a weak payoff matrix leads to decreasing cooperation levels. However, in dynamic networks just the opposite is observed. The reason is cooperating or defecting with a defecting neighbor makes no difference in a weak PD game. Players who are inclined to cooperate tend to break ties with all defective neighbors to try and get higher payoffs with others. This makes the network dynamic. Reputation was defined as a linear combination of the past action and the fraction of previous cooperative actions. They found significantly higher levels of cooperation were obtained when the reputation of their competitors was available.

These experimental results (and others) reveal some discrepancies between the models and human behavior. The Traulsen et al. [2010] study describe above is the same one mentioned previously where it was discovered humans use a selection intensity orders of magnitude higher than weak selection values. But the Traulsen human experiment had another significant finding: Nowak et al. [2004] posited that network reciprocity promotes cooperation but the Traulsen paper says maybe not so much. (See Helbing and Yu [2010] for a more detailed commentary.) What gets promoted is confusion when models and human experimental data don't agree.

Social dilemma models are rarely (if ever) validated. Usually in the literature validation is either just ignored or classified as "future work." It only leads to confusion when conclusions derived from one model conflict with the conclusions derived from another model. This conflict is not advancing the knowledge envelope. But human experimenters are not blameless either. Participants are almost always university undergraduates. This means the subjects are young, somewhat affluent, and educated. Most lack any real-world experience with taxes, utility bills, mortgages, and so forth. The social dilemma games also do not pose any real financial risk to the players. Indeed, most give the participants some small pay outs for their participation. That makes these experiments vastly different from real-world social dilemmas where mutual defection has financial consequences. Older, more experienced adults may choose differently. They are also more likely to be participants in those real-world social dilemmas.

But the point here is not to critique human experiment construction. The human experiments are what they are. And they do provide some data on how humans behave. There is no reason why researchers who hypothesize some new proximate cause for cooperation can't compare their model results to this data. In other words, researchers should start using the validation process described earlier. Ideally, this would develop into a full-blown arms race: human experimenters construct more realistic experiments with more diverse participant demographics and the game theorists develop more sophisticated models that imitate the experiments.

6.3 ASKING THE RIGHT QUESTIONS

There is a scene in the classic 1999 science fiction movie *The Matrix* that is relevant here. In that scene agents Smith and Brown are interrogating the human Morpheus. They are trying to get him to reveal the access codes to the Zion mainframe which will allow the machines to take over. Morpheus resists all of their interrogation methods. In frustration agent Smith asks why the serum isn't working. Agent Brown replies "perhaps we're asking the wrong questions." That reply may very well explain why current methods used by games researchers have produced so few answers regarding cooperation. Perhaps we too are asking the wrong questions.

The fundamental question social dilemma research is trying to answer is why do humans cooperate with non-kin. Indeed, this may well be the only question worth answering. Before proceeding further, it will be instructive to see exactly what questions are being answered by the social dilemma research community. The examples given below were selected because they contain examples of the type of analysis appearing in the literature.

1. **Example #1** (from Nowak et al. [2004])

   There are two strategies A and B. The strategy fitness is frequency dependent. Let i be the number of individuals using strategy A and let $w \in [0, 1]$ denote the selection intensity. Then the fitness of strategy A and B, respectively, is

   $$f_i = 1 - w + w[a(i - 1) + b(N - i)]/[N - 1]$$

   $$g_i = 1 - w + w[ci + d(N - i - 1)]/[N - 1],$$

   (6.2)

   where N is the population size and the payoff matrix for the row player is given by

   $$\begin{array}{cc} & \begin{array}{cc} C & D \end{array} \\ \begin{array}{c} \text{payoff to } C \\ \text{payoff to } D \end{array} & \begin{pmatrix} a & b \\ c & d \end{pmatrix}. \end{array}$$

   (6.3)

   Selection dynamics are a Moran process under frequency-dependent fitness with self play not permitted. At each time step an individual is chosen proportional to fitness. This individual reproduces and the offspring replaces a randomly chosen individual. Thus, N is constant.

   The probability of adding an A-offspring is

   $$\frac{i f_i}{i f_i + (N - i) g_i}.$$

   (6.4)

   At each time step the number of A individuals can increase by one, decrease by one, or stay the same. This can be represented as a Markov process with a tridiagonal transition matrix. Let $P_{i,i+1}$ represent the probability of increasing the number of A individuals by one. Then the Markov process defines a BD update rule with tridiagonal

$$P_{i,i+1} = \frac{i\,f_i}{i\,f_i + (N-i)g_i} \cdot \frac{N-i}{N}$$
$$P_{i,i-1} = \frac{(N-i)\,g_i}{i\,f_i + (N-i)g_i} \cdot \frac{i}{N} \tag{6.5}$$

and $P_{i,i} = 1 - P_{i,i+1} - P_{i,i-1}$. All other transition matrix entries equal zero.

The Markov process has two absorbing states[1]: $i = 0$ and $i = N$. Let x_i be the probability the population reaches the absorbing state $i = N$ when starting in state i. Then

$$x_i = P_{i,i+1}x_{i+1} + P_{i,i}x_i + P_{i,i-1}x_{i-1} \tag{6.6}$$

with boundary conditions $x_0 = 0$ and $x_N = 1$.

It can be shown that the probability ρ_A that a single individual A can invade a population of B players is given by

$$\rho_A = 1 \bigg/ \left(1 + \sum_{k=1}^{N-1} \prod_{i=1}^{k} \frac{g_i}{f_i} \right) \tag{6.7}$$

and selection favors A replacing B if $\rho_A > 1/N$.

2. **Example #2** (from Traulsen et al. [2007])

   In this work two individuals, A and B, are selected randomly. The probability A replaces B is given by the Fermi function

   $$p = \frac{1}{1 + e^{-\beta(\pi_A - \pi_B)}}, \tag{6.8}$$

   where $\beta/ge0$ is an inverse temperature from statistical fitness. The pairwise comparison process has an advantage over the Moran process because negative payoffs are allowed. The transition probability of changing the number of A players from j to $j \pm 1$ is

   $$P_j^{\pm} = \frac{j}{N} \frac{N-j}{N} \frac{1}{1 + e^{\mp\beta(\pi_A - \pi_B)}}. \tag{6.9}$$

   For weak selection ($\beta \ll 1$) the Fermi function can be expanded to make the transition probabilities

---

[1]An absorbing state is a state that once reached, it can never leave.

$$P_j^{\pm} \approx \frac{j}{N} \frac{N-j}{N} \left[\frac{1}{2} \pm \frac{\beta}{4} (\pi_A - \pi_B) \right].$$ (6.10)

Conversely, in the frequency-dependent Moran process the transition probabilities are

$$P_j^+ = \frac{j(1-w+w\pi_A)}{j(1-w+w\pi_A)+(N-j)(1-w+w\pi_B)} \cdot \frac{N-j}{N}$$

$$P_j^- = \frac{(N-j)(1-w+w\pi_B)}{j(1-w+w\pi_A)+(N-j)(1-w+w\pi_B)} \cdot \frac{j}{N},$$ (6.11)

where w is the selection intensity. Under weak selection, $w \ll 1$, this leads to

$$P_j^+ \approx \frac{j}{N} \frac{N-j}{N} \left[1 + w \frac{N-j}{N} (\pi_A - \pi_B) \right]$$

$$P_j^- \approx \frac{j}{N} \frac{N-j}{N} \left[1 - w \frac{j}{N} (\pi_A - \pi_B) \right].$$ (6.12)

The ratio $P_j^- / P_j * +$ is the same for both pairwise comparison and the Moran process. Under weak selection this ration can be approximated as

$$\frac{P_j^-}{P_j^+} \approx 1 - w (\pi_A - \pi_B).$$ (6.13)

This ratio is the same with $w \leftrightarrow \beta$. This ratio also determines the fixation probability ϕ_k where the population initially has k players choosing strategy A. Fixation refers to the state where the entire population chooses the same strategy, i.e., it "fixates" at strategy A.

Remark 6.1   Invasion is a fixation special case. Suppose everyone in a population chooses B and, through mutation, one individual switches to A. If eventually everyone switches to A, then A is said to "invade" the population. In other words, A invades B if the population goes from $k = 1$ to the absorbing state $k = N$. If there is no strategy A that can invade, then B is an evolutionary stable strategy (ESS). An ESS may be a NE but not vice versa (see Apaloo et al. [2014]).

Given initially k individuals choosing A, the fixation probability of A is

$$\phi_k = \frac{\sum_{i=0}^{k-1} \prod_{j=1}^{i} P_j^- / P_j^+}{\sum_{i=0}^{N-1} \prod_{j=1}^{i} P_j^- / P_j^+},$$ (6.14)

where P_j^+ is the probability of increasing the A players from j to $j+1$ and P_j^- the probability of decreasing j to $j-1$.

Fixation time refers to how long it take to reach fixation. Let t_k be the average time spent in transient states, i.e., starting from k initial players choosing A, transients states are where $1 \leq k \leq N-1$. The average time before reaching either fixation state of $k=0$ or $k=N$ is

$$t_k = 1 + P_k^+ t_{k+1} + \left(1 - P_k^+ - P_k^-\right) t_k + P_k^- t_{k-1}. \tag{6.15}$$

The authors investigated three different fixation times. Two were conditional fixation times: how long it takes to reach the state $k=0$ (everyone chooses B), and how long it takes to reach the state $k=N$ (everyone chooses A). The other was an unconditional fixation time: how long it takes to reach $k=0$ or $k=N$.

In an appendix they provided formal proofs on fixation times. For example, for the unconditional fixation time proof the payoff matrix was

$$\begin{array}{c} \\ A \\ B \end{array} \begin{array}{c} A \quad\; B \\ \begin{pmatrix} a_{11} & a_{12} \\ a_{21} & a_{22} \end{pmatrix} \end{array} \tag{6.16}$$

with $2u = a_{11} - a_{12} - a_{21} + a_{22} \neq 0$ and $2v = -a_{11} + a_{12}N - a_{21}N + a_{22}$, they defined $\chi_l = \exp[\beta l u + 2\beta v]$. Then

$$t_k = \phi_k S_n - S_k, \tag{6.17}$$

where

$$S_j = N^2 \sum_{n=1}^{j-1} \chi_{n+1}^{-n} \sum_{l=1}^{n} \frac{1 + \chi_{2l}^{-1}}{l(N-l)} \chi_{l+1}^l. \tag{6.18}$$

3. **Example #3** (from Kurokawa and Ihara [2009])

Groups of n individuals are randomly chosen out of a population of size N. Each group plays a general game with two strategies A and B. The payoff matrix is shown in Table 6.1. The columns indicate the number of players choosing A among the $n-1$ opponents. For example, if the focal player chooses A (B) and $n-3$ opponents chose A, then the focal player payoff is a_3 (b_3). The expected payoffs are

Table 6.1: Payoff matrix for the general game

	$n-1$	$n-2$	$n-3$	...	1	0
A	a_1	a_2	a_3	...	a_{n-1}	a_n
B	b_1	b_2	b_3	...	b_{n-1}	b_n

$$F_i = \sum_{k=1}^{n} \frac{\binom{i-1}{n-k}\binom{N-i}{k-1}}{\binom{N-1}{n-1}} a_k,$$

$$G_i = \sum_{k=1}^{n} \frac{\binom{i}{n-k}\binom{N-i-1}{k-1}}{\binom{N-1}{n-1}} b_k \tag{6.19}$$

and under weak selection $f_i = 1 - w + wF_i$ and $g_i = 1 - w + wG_i$. The fixation probability, ρ_A, of $i = 1$ players choosing A eventually reaching the absorbing state $i = N$ be given by Eq. (6.7). The fixation probability of B, ρ_B, can be determined similarly.

In a general n-player game with weak selection, the fixation probability of strategy A is given approximately by

$$\rho_A \approx \frac{1}{N} \frac{1}{1 - (\alpha N - \beta)w/n(n+1)}, \tag{6.20}$$

where

$$\alpha = \sum_{k=1}^{n} k\,(a_k - b_k)$$

$$\beta = -n^2 b_n + \sum_{k=1}^{n-1} k b_k + \sum_{k=1}^{n} (n+1-k)a_k. \tag{6.21}$$

Selection favors A if $\rho_A > 1/N$ and B if $\alpha N > \beta$. The ratio of the fixation probabilities shows the likelihood A will replace B and is given by

$$\frac{\rho_A}{\rho_B} \approx 1 + \frac{w}{n}\left[\gamma N - n(a_1 - b_n)\right] \tag{6.22}$$

with

$$\gamma = \sum_{k-1}^{n}(a_k - b_k). \tag{6.23}$$

The above examples are just a sampling of the elaborate calculations that commonly appear in the literature. Nevertheless, they represent a common them found in far too many papers: an obsession with calculating fixation probabilities and fixation times. These efforts are misguided. Does anybody really care about strategy fixation? And if so, why?

Such questions may seem somewhat surprising, even absurd, but they are nevertheless serious inquiries. In population genetics, fixation occurs when a gene pool contains at least two variants (alleles) of a particular gene, but eventually only one allele remains. In social dilemmas fixation means eventually the population evolves to a state where everyone uses the same strategy. Fixation at defection is not very interesting. Fixation at some non-defection strategy would however attract enormous interest. The conditions that produced such an outcome would be extensively debated as would the short-term and long-term consequences from a societal or psychological perspective. Well, if fixation is so important, where are the real-world social dilemma examples of a human group fixated at a strategy other than defection? If some do exist, then researchers should be modeling them instead of modeling prisoner's dilemma or snowdrift games. These example social dilemmas should be described in papers as a justification for deriving fixation probabilities and fixation times. If there are no examples—or there are, but so rare they could be considered outliers—then maybe fixation isn't so important after all.

IPD (or its N-player version, PGG) are the most investigated social dilemma games. A journal will not publish a paper unless it contains new ideas. This has led to researchers to introduce all sorts of PGG variants, each worthy of investigation and a potential source of future publications. Being different might get you published, but that doesn't mean it contributes anything useful to the body of knowledge. Some of these variants are rather odd and seem to be proposed primarily because they hadn't appeared in print before. Unfortunately, some of these papers have little (if any) relevance to human populations.

One example is the use of random payoff matrices. Eriksson and Lindgren [2001] studied 2-player games with fixed strategy choices. In each round an entirely new payoff matrix was randomly generated with entries sampled from the unit interval. Selection intensity was used and runs up to 60,000 rounds were conducted. Huang and Traulsen [2010] considered the case where a mutant strategy was introduced into a homogeneous population, but payoffs for the mutant are set by a payoff matrix whose entries are random variables with Gaussian distributions. This means the fixation probability itself now is a random variable. (Other distributions were also discussed.)

---

[1]One exception is survey papers, but they are not relevant to this discussion.

In social dilemmas people make choices and receive some payoff or return depending on the choices made by others. Thus, a player may make the same choice in two different rounds and wind up getting different returns in the two rounds. That makes sense. If one cooperates in a group, and the number of other cooperators changes, then the return should be different. Many social dilemma games follow this general paradigm and use frequency-dependent fitness to model payoff variation from round to round due to demographic changes. Using frequency-dependent fitness is a reasonable approach. On the other hand, making the payoff matrix entries random variables with some underlying distribution is not reasonable. Eriksson and Lindgren did not cite any actual social dilemmas that operate this way. Huang and Traulsen tried to justify their investigation by comparison with population genetics. But, as mentioned previously, comparing population genetics with human populations is not legitimate; the population dynamics are completely different.

Even more bizarre is adding noise to the payoff matrix. The idea is noise can model player errors where intended actions are not correctly executed. Noise can also model uncertainty about the payoffs. Zhang et al. [2016] investigated a prisoner's dilemma game on a square lattice with periodic boundary conditions. Cooperate and defect were the only strategies. Every round a player x engages in a pairwise PD game with his neighbors and accumulates a total payoff of p_x. This payoff is then converted into a fitness $F_x = (1 + \mu)p_x$ where $\mu = \alpha\chi$. $\alpha \in [0, 1]$ and χ is a random variable sampled on the interval [-1,1]. Player x updated his strategy by randomly choosing one neighbor y with fitness $F_y (= p_y)$ and adopting that strategy with probability $[1 + \exp([F_x - F_y]/K]^{-1}$ where K is user selected. Games lasted 2×10^5 rounds and cooperator frequency was recorded over the last 10^4 rounds. The authors concluded

> ... by introducing the element of noise into the measurement of fitness in the prisoner's dilemma game, our modified model remarkably facilitates the emergence of cooperation.

Essentially the authors argue uncertainty about payoffs leads to increased cooperation levels. Simulation results not withstanding, in the real world such a claim is preposterous. Who would continue to participate in a social dilemma if he continually is making errors or is clueless about potential returns? A previous chapter talked about commercial fisherman in the Tasman Sea off the coast of Australia, a quintessential example of a tragedy of the commons social dilemma. Any fisherman who is constantly making mistakes or has to guess about the payoffs will quickly go bankrupt. Zhang et al. believe it would lead to more cooperation among the fisherman.

But the major problem with much (most?) social dilemma computer modeling is in the analysis. To put it bluntly, researchers are asking the wrong questions. Many researchers are absolutely convinced fixation is an important and inherent property of every social dilemma game model. Consequently, they believe their peers will expect to see fixation derived, quantified, analyzed, or discussed in a paper—particularly if it's a journal paper. Model acceptance depends on it.

Algorithm 6.2 Analysis

1: Formulate a hypothesis about the proximate cause of cooperation
2: Design a model structure and parameter set to implement the hypothesis
3: Run simulations
4: Collect model results
5: Derive formulas for fixation probability, invasion probability fixation time, etc.
6: Use model results as formula arguments; Quantify fixation parameters and discuss.
7: (optional) Compare fixation probabilities and times with previously published findings
8: Exit

Algorithm 6.2 outlines the typical journal paper presentation of a social dilemma game. The first four steps are nearly identical in Algorithm 6.1. Where the two algorithms differ is after the fourth step. Model results are used in distinct ways which depends on the algorithm. In the validation algorithm model results are used to check congruence between the model and the physical system. Conversely, in the analysis algorithm model results are used to quantify model properties derived in the previous algorithm step. Any connection between model results and observed human behavior is deferred until later—if discussed at all.

This goes to the heart of the matter regarding social dilemma modeling: *a majority of researchers are analyzing the models rather than analyzing the model results.* Indeed, this is the main reason why the wrong questions are being asked. When you focus on the model, you ask questions about the model; when you focus on the results, you ask questions about the results.

Result-centric studies focus on explaining how results correspond to recorded data. Those who focus on the results will ask the following types of questions.

– Are the results consistent with data from actual social dilemmas or human experiments?

– Do the results reveal new proximate causes of cooperation or support existing ones?

– How can the model be adjusted to improve its accuracy?

– If the population reached fixation, does the data provide any explanations?

– Do strategy changes comport with expected human behavior?

– Do the results identify possible new areas of investigation?

Notice the common theme: these questions ask what the data tells us, not the mechanics of how the model produced those results. Fixation, for example, is important only if the data provides some explanation; if the model will fixate and how long it might take is of no concern. Result-centric studies worry about whether the results are believable and what we can learn about cooperation from them.

In sharp contrast, model-centric studies focus on explaining how the model behaves. Those who focus on the results will ask the following types of questions.

- What caused the model to fixate?

- How is the equation for fixation probability derived? for fixation time?

- What is the probability of a successful invasion?

- Can weak selection help develop approximation formulas?

- How do the results change when selection intensity is varied?

- How would the results differ if the strategy update rules are changed?

- How does network node degree affect the game results?

- Will changing the intensity of interaction between players affect cooperation levels?

All computer models have adjustable parameters. Model-centric researchers concentrate on studying how tweaking various parameter values might affect the model's predicted cooperation levels. Effectively this is nothing more than a parameter sensitivity analysis. Moreover, it is the reason behind the growth in PGG variants: each new parameter added to a model is a new model to investigate.

The Zhang et al. [2016] work described previously is a perfect example. The payoff matrix was

$$\begin{array}{cc} & \begin{array}{cc} C & D \end{array} \\ \begin{array}{c} \text{payoff to } C \\ \text{payoff to } D \end{array} & \begin{pmatrix} 1 & 0 \\ b & 0 \end{pmatrix}, \end{array}$$

where $1 < b < 2$. Their model, in addition to b, also had a parameter $\alpha \in [0, 1]$ controlling the degree of noise in the fitness mapping. The analysis concentrated on determining how the co-operator frequency depends on b and α. As another example, Niu et al. [2018] studied an IPD game with cooperation, defection or loner as the three strategy choices. (Loners do not interact with others but received a fixed return σ each round instead.) Players were arranged on a 200×200 square lattice with periodic boundary conditions. What is unusual about their work is players played against themselves too. A cooperator could receive an additional reward $0 < \delta < 1$ by cooperating with himself. Not surprisingly, the authors found that

Compared to the traditional version ($\delta = 0$), having additional reward δ provides an advantageous environment for cooperators to survive, where cooperative behavior is markedly enhanced when temptation to defect b [sic] less than its threshold.

It is not intuitively obvious why a player would defect against himself. The authors provided no explanation.

Parameter sensitivity studies tell us a lot about a particular model—but unless those parameters have clear connections to aspects of human behavior, they tell us nothing about cooperation. Indeed, depending on the model's construction, it may be possible to justify any cooperation hypothesis if one can find a suitable parameter value.

Fixation probability and time may not necessarily be under control of a specific parameter value, but it is influenced by the model's evolutionary process controlling strategy changes. Just like adding a new parameter to an existing model creates a new model to investigate, so does changing the evolutionary process. Some investigations produce results that frankly are not very helpful. For example, Liu et al. [2017] determined:

> *For the prisoner's dilemma game and the coordination game, the fixation probability of a single cooperator in our mixed process is higher than that in standard Moran process under weak selection. However, the fixation probabilities under the coexistence game are lower than that in standard Moran process under weak selection.*

It is not clear how this finding provides any insight into human cooperation. It seems continually studying fixation probability or time is unlikely to provide much insight either. For example, Traulsen et al. [2006] studied IPD games and found:

> *In spite of the exponential increase of the fixation probability of cooperators with the initial number k, only for very weak selection ($\beta = 0.01$) do cooperators acquire reasonable chances in a population as small as $N = 20$.*

As pointed out in a previous chapter, weak selection results do not hold at higher selection intensity levels so the usefulness of this type of information is doubtful. Indeed, it is reasonable to question the validity of any fixation equations since they are almost always derived assuming weak selection. But there is a more important point here. Ultimately, it may not make much sense to study fixation unless its promoters provide evidence it regularly takes place in real-world social dilemmas. If it does not, it is hard to defend why it is so important and why it deserves so much attention.

Unfortunately, studying the model itself remains the emphasis of far too many papers. This situation must change if there is any hope of making progress. Constructing a model and then deriving fixation equations isn't helpful, and adds nothing to our understanding about cooperation, if fixation isn't prevalent. Likewise, any claims about what does or does not promote cooperation in humans are worthless if they are based on weak selection modeling results.

New models should never be introduced because "hey, nobody tried this before." Only models that generate predictions consistent with real-world data should appear in print. Validation compares model predictions with real world data. Just like prices help business concentrate resources on what customers want, validation helps concentrate research efforts on what is likely to produce new insight and reveal new avenues of discovery. Without validation of some kind it is impossible to know how—or even if—any of the predictions obtained from a model apply in human populations. Validation efforts must have greater importance. More human experiments should be conducted and better dissemination of real-world social dilemma data is needed. Unless this is done, we may continue to ask the wrong questions.

6.4   SUMMARY

- Many papers concentrate on analyzing a model's mathematical behavior. This takes precedence over interpreting a model's predictions.

- Deriving a model's fixation time and fixation probability is a major topic in many studies. It is not obvious why fixation is so important since examples of fixation in human populations are lacking.

- Models with random payoff matrices do not mimic human behavior in social dilemmas.

- Model results are seldom compared with human experimental data.

- Model validation has not received anywhere near the attention it deserves.

- Unvalidated model results cannot provide meaningful insight into human cooperation.

CHAPTER 7

The Path Forward

In this chapter a review of the major problems with current computer modeling practices will be revisited, however sometimes with a different perspective. This will be followed by several recommendations.

7.1   THE MORAN PROCESS

The Moran process randomly chooses two individuals out of the population, one proportional to fitness (accumulated payoffs) and the other with equal probability. The strategy of the first individual replaces the strategy of the second individual with some probability. The broad appeal of the Moran process is understandable: it is easy to understand and trivial to program.

But as pointed out previously, it does not match reality. Mathematical convenience and ease in programming are not sufficient justifications for its widespread use. While it does provide a mechanism for evolving strategy changes over time, it doesn't adequately describe human behavior. The strategy changes are mandated. Not only is a player forced to use a new strategy, but he has no input as to what that new strategy will be. Any player would reasonably object to such rules and, indeed, may even decline to participate. People want to make their own decisions—particularly if there are financial payoffs at stake. That said, is there then some other plausible reason for using a Moran process?

A paper relevant to this discussion was recently published by Hilbe et al. [2018]. Their paper described research on a stochastic game in which strategies evolved through a Moran process. The details of the stochastic game itself are not important here. What is important was the following statement that appeared:

> *Traditionally, work on stochastic games considers rational players who can employ arbitrarily complex strategies, but does not focus on the dynamics of how players adapt their strategies. We introduce an evolutionary perspective to stochastic games. Players do not need to act rationally, but instead they experiment with available strategies and imitate others depending on success.*

That quote is intriguing for a couple of reasons. The evolutionary perspective introduced was strategy changes via a Moran-type process. (Two individuals were randomly chosen, a "learner" and a "role model." π_L and π_R represent their fitness. The learner adopts the strategy of the role model with probability $p = 1/[1 + e^{-\beta(\pi_R - \pi_L)}]$ where $\beta \geq 0$ is the selection intensity.) The fact traditionally stochastic game research hasn't focused on the dynamics of strategy adaptation is a valid criticism. People study social dilemma games to identify conditions that promote cooperation. Player strategies ultimately determine the cooperation levels. Therefore, the reasons why players change strategies is always pertinent. That said, given the various issues associated with its use, the fact historically strategy adaptation dynamics have been ignored in stochastic games is not a compelling reason for using a Moran process. But what is more interesting—and apparently another reason for using a Moran Process—is the implication experimenting with available strategies and imitating successful ones somehow constitutes irrational behavior.

What is rational behavior in a social dilemma? At this point we will avoid going off into the weeds discussing the principles of rational choice theory. But before answering this question it is important clear up the difference between two terms that are sometimes conflated.

Definition 7.1 (payoff) *A reward or benefit received after interacting with one or more other players.*

Payoffs have tangible value. For example, a player may receive $49 after a public goods game round. The $49 is a payoff.

Definition 7.2 (utility) *The level of satisfaction associated with receiving a payoff.*

The utility of getting a $49 return can be high or low depending on what return the player anticipated getting back.

Utility is a qualitative measure whereas payoff is quantitative. They are not synonymous terms. In fact, maximum utility doesn't necessarily mean maximum payoffs. Although most researchers use payoffs, a switch to utilities may be more beneficial in the future. Utilities, being qualitative, are better suited for examining motives behind strategy changes.

Homo economicus or the "economic man" is strictly rational. Possessing complete information about the economy, and the consequences of any actions, he can optimally pursue wealth for his own self-interest. This is the common definition of rationality used in a majority of social dilemma games. That is, a rational player will always choose a strategy most likely to maximize his payoff. Critics of *homo economicus*, such as the Nobel Prize winning economist Friedrich Hayek, argue humans do behave irrationally. Rational behavior stems from perfect knowledge of all relevant facts in a given situation. These special circumstances are unlikely to occur due to the complexities of society. The important point here though is the critics do not interpret irrational behavior as random behavior; they only mean people don't always act in their own self interest.

On the other hand, Hilbe et al. [2018] imply irrational behavior is unpredictable and therefore using a random process, like a Moran process, is a legitimate way of picking strategies. Strategies can be picked at random, tested for efficacy, and either copied or rejected depending on success. But does experimenting with different strategies and then choosing the best performing constitute irrational behavior?

Experimenting with different strategies is not irrational behavior per se. A player may not be able to actually try out each strategy before picking one but nothing precludes doing strategy trade-offs through thought experiments. Hilbe et al. [2018] imply strategy experimentation is done haphazardly, choosing strategies for comparison at random via some coin flipping process. Rational players compare the current strategy against all available alternatives, not against just one randomly selected as in done in a Moran process. Payoff difference is an important criteria for comparison, but not the only one. Confusing or difficult to implement strategies should be quickly discarded even if they have higher payoffs. Thus, it is not the experimentation itself that is irrational, but how it is conducted. It is the Moran process that exhibits irrational behavior because of how it does strategy replacement.

For our purposes the notion of rationality used by most game theorists—i.e., an individual who always acts in his own self interest—is a sufficient working definition. A rational decision then is a deliberate, intentional act which is the outcome of some thought process. However, under this definition "rational" should not be conflated with "logical." One may argue a decision was not properly thought out, or naive, or did not consider certain facts. Those arguments question whether the decision was wise. They do not mean the decision wasn't deliberate or lacked some cognitive process. That said, it is still important to clarify what is meant by "in his own self interest." Self interest is obvious with payoffs, but what about with utilities?

Consider a PGG with three strategies: cooperator, defector and punisher. Punishers act like cooperators except they impose a penalty on defectors. This penalty lowers the defector's payoff and is intended to convince the defector to start cooperating. The punisher pays a cost to impose this penalty. Thus, punishers get a lower return (payoff) than cooperators. If self interest is defined in terms of payoffs, then switching from a cooperator to a punisher makes no sense. Or does it? Consider if there were no punishers. Then there is no incentive for defectors to stop defecting. Indeed, the higher payoffs may induce others to stop cooperating, particularly if defectors are in the minority. So the cooperator contemplating switching to a punisher has a choice: don't punish and let defectors grow or accept a lower return by paying the cost to punish. But suppose the punishment is successful and defectors start cooperating. The higher contributions will benefit the punisher in the future Thus, punishment is a rational choice because taking a lower return in the short term can lead to greater returns in the long term. Self interest based on payoffs makes sense.

Suppose self interest in this PGG is based on utilities. Human experiments have shown emotions—particularly anger—is the motivation behind the decision to punish (e.g., see Parks et al. [2002]). Most people consider emotional responses to be irrational responses. But in this case punishment is justified using utilities to define self interest. The cooperator decides to punish as a retaliation against what is perceived to be unfair behavior. Thus, an emotional response can be rational. As before, short-term loss can result in long term gains.

So if a Moran process is a poor method for evolving strategies in a population, what other alternatives are there? One possibility is fuzzy systems. Fuzzy logic attempts to mimic the human decision-making process. It provides a means of handling imprecise or vague information. With fuzzy logic each player independently make decisions during each round of the game. No random processes are involved either.

Greenwood et al. [2019] used fuzzy logic in a PGG. Defectors who don't contribute are first-order free riders while cooperators who don't punish are second-order free riders. (Punishers pay a cost to impose the punishment so they contribute, but get a lower return than cooperators.) In this PGG cooperators, once they are in the minority, all agree to start punishing defectors. The punishment a defector gets depends on the number of punishers. Hence, it is important that all punishers remain *trustworthy* —i.e., they initially agreed to punish and they will continue to punish. There is always a temptation to reduce costs so some punishers may decide to just go back to contributing to avoid paying the punishment cost. Players who originally agreed to punish, but later switch back to just cooperators are *untrustworthy* . The fuzzy system determined when the remaining punishers would start punishing these untrustworthy, second-order free riders. Table 7.1 shows the strategies and their payoffs.

Table 7.1: Payoffs for different strategy types

Strategy	Punishes	Payoff
C_1	None	$rcn/N - c$
C_2	1st order free rider	$rcn/N - c - \eta$
C_3	1st, 2nd order free rider	$rcn/N - c - \eta - \gamma$
D	–	$rcn/N - \beta$

Here c is the contribution to the pool, r is the pool multiplication factor, n the number of cooperators, and N the population size. The punishment costs are fixed with $\gamma = \eta < c$. Once in the minority the cooperators (C_1) switch to punishers (C_2). The punishment β linearly increases with the number of punishers. If there are initially M punishers then defectors get a punishment $\beta = M\alpha$. α is chosen large enough so that for the M value the punishment will be effective. However, if M decreases, it will no longer be effective and defectors will increase in the population.

Two strategy frequencies are important. x is the frequency of untrustworthy punishers out of the M players who initially agreed to punish. For example, suppose initially ten cooperators decided to punish defectors. If all actually punish, then $x = 0.0$, if later 3 of them stop punishing, then $x = 0.3$. y is the frequency of defectors in the population. Note $x + y \neq 1$. (We can let $z = 1 - y$ be the cooperator + punisher frequency in the population.)

All fuzzy systems have membership functions and a rule base. Membership functions determine the degree to which a variable belongs to some category. Figure 7.1 shows the membership functions. Suppose $x = 0.4$. Then the membership of x in the fuzzy set ZE (zero) is 0, approximately 0.6 in fuzzy set LO and about 0.2 in fuzzy set HI.

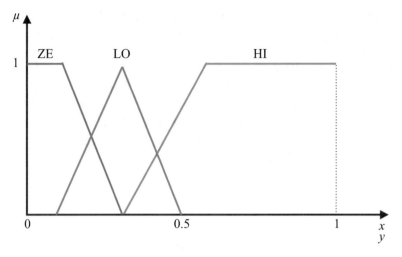

Figure 7.1: Membership functions for the fuzzy variables. The domain of discourse is the unit interval because both x and y are frequencies (see text).

Table 7.2: Fuzzy rule base

Number	Rule
1	If x is $\neg$ZE and y is HI then $f_1(x, y) = \frac{4}{5}x + y$
2	If x is ZE and y is ZE then $f_2(x, y) = 5x + 2y$
3	If x is HI and y is LO then $f_3(x, y) = 4x + 4y$

Fuzzy rules are if-then type of rules. The antecedent is the "if" part and the consequence is the "then" part. The fuzzy rule base is shown in Table 7.2. The consequent in each rule is a function $f(x, y) = ax + by$ where a and b depend on the rule. Each rule then is defuzzified to produce a crisp output θ where

$$\theta = \frac{\sum_{j=1}^{3} w_j f_j(x, y)}{\sum_{j=1}^{3} w_j}. \tag{7.1}$$

This crisp output is the penalty imposed on the untrustworthy second-order free riders. An example show how the inference weights (w_j) are calculated. In the first rule the two antecedent terms "x is $\neg$ZE" and "y is HI" have membership values indicating the degree of membership in the indicated fuzzy set (see Figure 7.1). Let these membership values be m_1 and m_2, respectively. Then $w_1 = \min(m_1, m_2)$. To calculate m_1 first compute the membership value v for the set ZE and then $m_1 = 1 - v$. The consequent of this rule is $f_1(x, y) = 4x + y$.

Figure 7.2 shows how θ varies with the frequency of untrustworthy players x. Three regions of total cooperation frequency are shown: $z = 0.4$ where cooperators are still in the minority; $z = 0.6$ where they have a small majority; and $z = 0.8$ where they have a strong majority. The rules, presented in Table 7.2, were specifically constructed to produce this type of response and indicates different levels of tolerance for second-order free riders depending on the total cooperation frequency. When cooperators have a small population majority (green plot) up to about $x = 0.3$ the second-order free rider punishment is only moderate. However, when the total cooperation frequency is $z = 0.8$—i.e., when most of the defectors are purged—there is little tolerance for free riding. Notice the punishment for untrustworthy players is severe when $x \approx 0.11$.

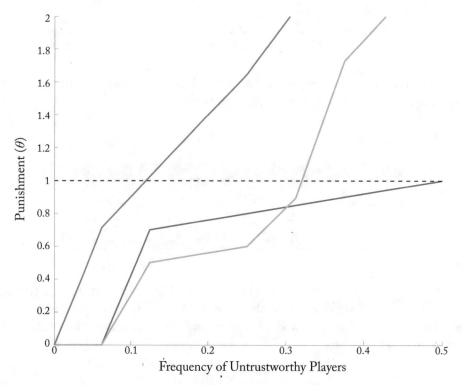

Figure 7.2: Punishment (θ) vs. untrustworthy player frequency when the total cooperator frequency in the population is 0.4 (blue), 0.6 (green), and 0.8 (red). Dashed line shows the cost to punish second-order free riders when $c = 1$.

To test this fuzzy system a schedule was constructed specifying x increases. One player became untrustworthy at iteration (round) 2, four more at iteration 4 and one player at iteration 11. Figure 7.3 shows how the total cooperation frequency changes when there is no punishment, defector punishment only, and defector plus untrustworthy player punishment. Discrete

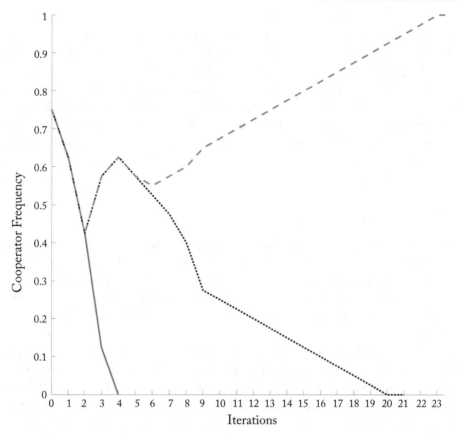

Figure 7.3: Cooperator frequency vs. iterations for no punishment (blue), first-order free rider punishment (black) and both first- and second-order free rider punishment (red).

replicators determined how the strategies evolved. Defectors quickly dominate when there is no punishment as shown by the blue plot. Notice that punishing defectors only is not sufficient. In the black plot punishment is initially effective causing an increase in cooperator frequency. However, at iteration 4 the number of untrustworthy player increased, thereby dropping the punishment level of defectors. Defectors starting increasing until ultimately all players defected. Only when the fuzzy system was brought online so both defectors and untrustworthy players were punished did cooperators prevail.

The rule base contains different rules reflecting different tolerances for second-order free riding depending on the total cooperator frequency. In this particular research all trustworthy players used the same rule base. However, individuals may have different levels of tolerance for free riding. Nothing precludes each player having its own unique set of rules. Moreover, each player could have a different set of membership functions (although that is rarely done). There are no probabilities and no sampling, so a single simulation produces results that are repeatable.

The games community has largely ignored fuzzy logic methods for adapting social dilemma strategies. Fuzzy logic is an effective and flexible method that is tunable to specific social dilemma scenarios. There are no random processes involved so it is unnecessary to make multiple simulations and then average results. It also is a perfect framework for agent-based modeling. Replicator dynamics take a holistic view of strategy evolution. Conversely, agent-based approaches are far more realistic because each agent acts independently, just like individual participants do in a human population. Fuzzy logic approaches are a viable alternative to Moran process game dynamics.

7.2   SPATIAL GAMES

Many field studies have provided conclusive evidence that in all societies people tend to interact with relatively few other people. This characteristic holds even in large populations. Spatial games attempt to emulate this natural inclination by forming neighborhoods (Von Neumann or Moore) and then restricting interactions to neighbors. The problem is these neighborhoods are not just artificial in structure, but they can easily produce erroneous results leading to faulty conclusions.

Normally a player gets a payoff only when he is the focal player interacting with his neighborhood. Figure 7.4 depicts a partial spatial game with Von Neumann neighborhoods. The neighborhood of A is $\{1, 3, 5, B\}$ and the neighborhood of B is $\{A, 2, 6, 4\}$. Suppose after a round of a PGG player A wants to imitate the strategy of B because B has a higher payoff. This would maybe make sense if A and B only interacted with the same players. But player B's payoff is mostly acquired from interactions with $\{2, 6, 4\}$—players A never interacts with. A far better approach is to stop using spatial games altogether and instead start using *tag-mediated games*.

An analogy will help illustrate the concept. At any football game it is common to see spectators wearing hats or t-shirts with the logo of the home team. Two fans wearing the home team's apparel may nod or smile at each other while standing in line at the concessions stand. They may not actually know each other, but they recognize they have something in common. This is the rationale behind tag mediation. Each player is assigned a color. Players interact each round only with other players with the same color (tag). Greenwood [2013] identified several advantages tag-medicated games have over spatial and network games.

1. *Modeling human experiments*

   In some human experiments the neighbors were fixed while in others they changed every round. Agents can only interact with other agents with the same tag. Fixed neighborhoods are created by assigning the tags and never changing them. Random shuffling of neighbors is accomplished by randomly reassigning tags each round. Tag-mediation can thus create models that mimic human experiments.

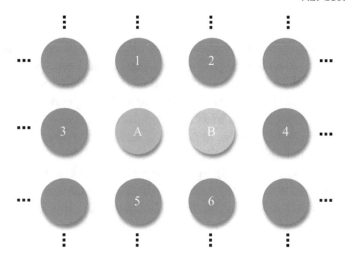

Figure 7.4: A partial spatial game with Von Neumann neighborhoods.

2. *Different group sizes*

   Spatial games artificially create neighborhoods with the same size. Humans don't normally do this even in large populations. Suppose the color tags are {red, green, yellow, white, and blue}. Each group is assigned one of the five colors. It is now possible, say in a population of size $N = 100$, to have group sizes of 10, 50, 4, 16, and 20. This is not possible in a game with players arranged on a 10×10 lattice.

3. *Emulating network structures*

   Humans do not naturally arrange themselves in lattices but they sometimes do associate in either random networks or scale-free networks. Random networks can be created randomly assigning the tags. A scale-free network is created by making a few tag groups very large while making most of the other tag groups relatively small.

4. *Dynamic networks*

   Dynamic networks are created by adapting the interaction probabilities. In tag-mediated games this is easily accomplished by periodically shuffling the tags. Group sizes can be probabilistically altered at the same time.

   Tag-mediated games overcome the artificial interactions inherent in spatial games. They are easy to implement. Neighborhood sizes are no longer constrained to be of equal size and group memberships can be fluid. Overall they provide far more flexibility than spatial games.

7.3 POPULATION SIZES

Continuous replicator equations are frequently used to predict strategy evolution. The problem is they require infinite populations sizes. For many years it has been known finite populations can have considerably different dynamics than infinite size populations (e.g., see Fogel et al. [1998], Nowak and Sigmund [2004]). Any conclusions about cooperation made using an infinite population may therefore not be valid in a finite population. It is also worth noting there is no such thing as an infinite population size in nature. Consequently, there is no good rationale for studying an infinite population.

Researchers need to stop studying infinite population size games. This mean continuous replicator equations should no longer be used to predict strategy evolution. Of course there is no problem with using discrete replicator equations. The population size issue is discussed further in the next section.

7.4 MODEL VALIDATION

Both validation and verification play important roles is building a correct model from which we can formulate correct conclusions [Sargent, 2011]. Figure 7.5 shows how these two apply. The conceptual model captures the dynamics of the real-world social dilemma. It mathematically describes the dilemma's dynamics and can be formulated in a variety of different ways such as replicator equations or with a set of interacting agents. Verification is a process that checks if the computer model was correctly programmed. That depends on the researcher's programming skills and requires sufficient testing. Here our focus is on the validation, which is more difficult.

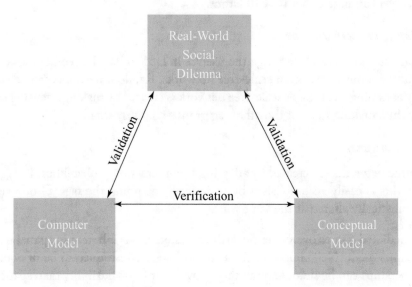

Figure 7.5: The validation/verification process.

Remark 7.3

Verification and validation are sometimes conflated. Basically, they ask different questions. The difference is in the ordering of two words.

Verification: Did we build the **thing right**?
Validation: Did we build the **right thing**?

Validation begins after the computer model is verified. The social dilemma environment is duplicated in the computer model. Observations from the social dilemma are then compared against the computer model's outputs. The comparison test produces either a "yes" or a "no" answer. If they are in some sense ϵ-close, then the answer is "yes" and the model is validated; otherwise the answer is "no." A no answer requires reevaluating the conceptual model which, in turn, will require re-verification of the computer model. Indeed, any conceptual model change requires re-verification of the computer model. These steps are repeated until the computer model results and the social dilemma observations are ϵ-close. Validation proves the hypothesis.

Multiple validation tests are required whenever there are more than one social dilemma scenarios of interest. The computer model is considered fully validated only after all validation tests pass. If the computer model is fully validated, then the conceptual model is also fully validated. In other words, a verified computer model correctly implements the conceptual model, and a validated computer model validates the conceptual model. Once the conceptual model is fully validated, any hypothesis about conditions that promote cooperation can be incorporated into it. The computer model is verified and then can be used to make predictions about the real-world social dilemma's behavior.

It is unreasonable to require, let alone expect, researchers conduct human experiments to support any claims about cooperation resulting from computer model simulation analysis. Nevertheless, there are some things that can be done to facilitate future validation efforts. Most conceptual models are designed with some social dilemma in mind, but unfortunately little thought is often given to the model structure itself. Poorly conceived structures can't be validated.

For example, one important thing is to stop using unrealistic model population sizes. In the previous section infinite populations sizes were discussed. The continuous replicator equations, which use infinite populations, do have some advantages: they are easy to get up and running plus they do give some concrete idea of whether a proposed hypothesis on cooperation shows promise. Researchers just need to remember that the finite population may show somewhat different results. Obviously, an infinite population model cannot be validated. But finite population models will have the same problem if the populations are too large.

Social dilemmas come all sizes from small groups to society at large. The greatest insight into cooperation among unrelated humans will most likely come from studying behavior in small to medium size groups or organizations. Small populations are much easier to manipulate making it much easier to set up the conditions for testing a hypothesis. It is also easier to collect the results and conduct post-event surveys to ascertain why people made particular decisions at particular times.

Regardless of what claims about cooperation arise from computer simulations, eventually they need to be tested. Spatial games described in the literature are always 100×100 or larger ($N \geq 10,000$). Such large size population size choices seem arbitrary and are never justified by the authors. It is worth noting none of these models were validated—and never will be either.

Consider the logistics required to validate any spatial game listed in Table 3.1. First, thousands of candidates would have to be identified, contacted and agree to participate. Next, the infrastructure for running the game needs to be defined and installed. Administering any game with 10,000 or more players will be, to put it euphemistically, challenging. Results have to be collected in a timely manner after each round. Most human experiments use undergraduate college students. Players normally can keep any accumulated payoffs or sometimes players are offered some payment ($\approx$ €20–40) as an incentive to participate. Even a small research budget could handle payments for an 8×8 game, but payments for a 100×100 game would be prohibitively expensive.

The 100×100 game seems to be the default size for spatial games. Unfortunately, that is too big to ever validate, which makes predictions about cooperation coming from that model questionable at best. But there is a simple solution to this problem. Any credible predictions about cooperation should not be overly sensitive to population size. Instead of simulating a 100×100 spatial game, simulate a 7×7 or 8×8 game. That size is small enough to validate. Besides, it takes the same amount of time to program a 7×7 game as it does a 100×100 game. Therefore, there is no need to use default sizes in spatial games.

Remark 7.4   Laboratory experiments with social dilemma games are an ideal setting for studying human behavior. Unfortunately, it is the researchers that speculate on player motives rather than asking the players to explain their strategy choices. (The laboratory experiment by Seip et al. [2014] is a notable exception.) The games community and the social science fields would be better served if post-game surveys were always conducted. This would also help guide the construction of future human experiments.

This leads us to the crux of the problem with social dilemma game modeling: too many models are unrealistic. Weak selection results don't hold at higher selection levels and humans use those higher levels. What value then are model results derived under weak selection? Humans don't make strategy choices using anything remotely close to a Moran process. How then do results from a model using a Moran process add to our understanding of human behavior in social dilemmas? Spatial models with populations in the thousands are never going to be validated. Can a hypothesis actually be affirmed using a model that cannot be validated?

These circumstances must change if we hope to make any progress. Weak selection assumptions should be abandoned. Any claims about cooperation obtained from models that are impossible to validate add little to our understanding. Population sizes must be scaled back to levels where model validation becomes feasible. Predictions coming from models that cannot be validated contribute little if anything. In fact, validation efforts demand much greater emphasis. Maybe it is time for the games community to admit the Moran process has now been fully exploited and it is too simple to provide any further benefits. A deeper understanding of human cooperation may only come from alternative approaches for choosing strategies which may require more involvement by behavioral economists. Until changes like these begin, perhaps we will continue to ask the wrong questions.

7.5   SUMMARY

- Rational players are self-interested. Players sometimes purposely exhibit irrational behavior in the short term as a strategy. However, short-term irrational behavior does not justify using a Moran process.

- The Moran process may be too simplistic to produce a deep understanding of human cooperation. Alternative approaches to control strategy changes are needed to replace the Moran process. Fuzzy systems are one possibility.

- Tag-mediated games are a viable alternative to spatial and network games.

- A hypothesis about cooperation is proven by validating the model that tested it.

- A population that is too large makes it impossible to validate the computer model. Virtually all spatial games described in the literature cannot be validated due to their population size.

- Predictions from models that can't be validated contribute nothing meaningful to our understanding of cooperation in human groups. Models must be designed to facilitate validation efforts.

- Model validation has not received anywhere near the attention it deserves.

D. Rand, M. Nowak, J. Fowler, and N. Christakis. Static network structure can stabilize human cooperation. *Proc. National Academy of Sciences*, 111(48):17093–17098, 2014. DOI: 10.1073/pnas.1400406111 44

Z. Rong, Z. Wu, and G. Chen. Coevolution of strategy-selection time scale and cooperation in spatial prisoner's dilemma game. *EPL*, 102:68005, 2013. DOI: 10.1209/0295-5075/102/68005 20

C. Sample and B. Allen. The limits of weak selection and large population size in evolutionary game theory. *Journal of Mathematical Biology*, 75: 1285–1317, 2017. DOI: 10.1007/s00285-017-1119-4 27

F. Santos and J. Pacheco. Scale-free networks provide a unifying framework for the emergence of cooperation. *Physical Review Letters*, 95:098104, 2005. DOI: 10.1103/physrevlett.95.098104 17

R. Sargent. Verification and validation of simulation models. In S. Jain, R. Creasey, J. Himmelspack, K. White, and M. Fu, Eds., *Proc. Winter Simulation Conference*, pages 166–183, 2011. DOI: 10.1109/wsc.1994.717077 66

E. Seip, W. Van Dijk, and M. Rotteveel. Anger motivates costly punishment of unfair behavior. *Motivation and Emotion*, 38:578–588, 2014. DOI: 10.1007/s11031-014-9395-4 4, 27, 68

J. M. Smith. The theory of games and the evolution of animal conflicts. *Journal of Theoretical Biology*, 47(209–221), 1974. DOI: 10.1016/0022-5193(74)90110-6 8

A. Stivala, Y. Kashima, and M. Kirley. Culture and cooperation in a spatial public goods game. *Phyical Review E*, 94:032303, 2016. DOI: 10.1103/physreve.94.032303 20

G. Szabó and C. Toke. Evolutionary prisoner's dilemma game on a square lattice. *Physical Review E*, 58(1):69–73, 1998. DOI: 10.1103/physreve.58.69 20

A. Szolnoki, J. Vukov, and G. Szabó. Selection of noise level in strategy adoption for spatial social dilemmas. *Physical Review E*, 80(5):056112, 2009. DOI: 10.1103/physreve.80.056112 20

J. Tanimoto. The impact of initial cooperation fraction on the evolutionary fate in a spatial prisoner's dilemma game. *Applied Mathematics and Computation*, 263:171–188, 2015. DOI: 10.1016/j.amc.2015.04.043 20

P. Taylor, T. Day, and G. Wild. Evolution of cooperation in a finite homogeneous graph. *Nature*, 447:469–472, 2007. DOI: 10.1038/nature05784 20, 28

A. Traulsen. Mathematics of kin and group selection: Formally equivalent? *Evolution*, 64(2):316–323, 2010. DOI: 10.1111/j.1558-5646.2009.00899.x 27

A. Traulsen, M. Nowak, and J. Pacheco. Stochastic dynamics of invasion and fixation. *Physical Review E*, 74:011909, 2006. DOI: 10.1103/physreve.74.011909 42, 55

A. Traulsen, J. Pacheco, and M. Nowak. Pairwise comparison and selection temperature in evolutionary game dynamics. *Journal of Theoretical Biology*, 246:522–529, 2007. DOI: 10.1016/j.jtbi.2007.01.002 47

A. Traulsen, D. Semmann, R. Sommerfeld, H. Krambeck, and M. Milinski. Human strategy updating in evolutionary games. *Proc. National Academy of Sciences*, 107(7):2962–2966, 2010. DOI: 10.1073/pnas.0912515107 29, 44, 45

H. Wang, Y. Sun, L. Zheng, W. Du, and Y. Li. The public goods game on scale-free networks with heterogeneous investment. *Physica A: Statistical Mechanics and its Applications*, 509:396–404, 2018. DOI: 10.1016/j.physa.2018.06.033 20

T. Weber, O. Weisel, and S. Gächter. Dispositional free riders do not free ride on punishment. *Nature Communications*, 9(1):2390, 2018. DOI: 10.1038/s41467-018-04775-8 27

G. Wild and A. Traulsen. The different limits of weak selection and the evolutionary dynamics of finite populations. *Journal of Theoretical Biology*, 247:382–390, 2007. DOI: 10.1016/j.jtbi.2007.03.015 26

B. Wu, P. Altrock, L. Wang, and A. Traulsen. Universality of weak selection. *Physical Review E*, 82:046106, 2010. DOI: 10.1103/physreve.82.046106 29

B. Wu, J. Garcia, C. Hauert, and A. Traulsen. Extrapolating weak selection in evolutionary games. *PloS*, 9(12):e1003381, 2013. DOI: 10.1371/journal.pcbi.1003381 28, 29

Y. Wu, S. Zhang, and Z. Zhang. Environment-based preference selection promotes cooperation in spatial prisoner's dilemma game. *Science Reports*, 8:15616, 2018. DOI: 10.1038/s41598-018-34116-0 20

M. Wubben, D. De Cremer, and E. Van Dijk. How emotion communication guides reciprocity: Establishing cooperation through disappointment and anger. *Journal of Experimental Social Psychology*, 45(987–990), 2009. DOI: 10.1016/j.jesp.2009.04.010 27

G. Zhang, T. Hu, and Z. Yu. An improved fitness evaluation mechanism with noise in prisoner's dilemma game. *Applied Mathematics and Computation*, 276:31–36, 2016. DOI: 10.1016/j.amc.2015.12.013 52, 54

Author's Biography

GARRISON W. GREENWOOD

Garrison W. Greenwood received a Ph.D. in Electrical Engineering at the University of Washington in Seattle, WA. He spent over a decade in industry designing multi-processor embedded systems and computer models for companies including Boeing, Honeywell, and Space Labs Medical. He then entered academia where he is currently a professor in the Electrical and Computer Engineering Department at Portland State University, Portland, OR. He was a Visiting Faculty at the University of New South Wales, Canberra, Australia for all of 2013. His research interests are evolvable hardware, cyber-physical systems, and evolutionary game theory.

Dr. Greenwood has been actively involved in the IEEE Computational Intelligence Society (CIS). He is the past chair of the Evolutionary Computation Technical Committee and served four years as the CIS Vice-President of Conferences. From 2006–2014 he was the Editor-in-Chief of the *IEEE Transactions on Evolutionary Computation*. He is currently serving on the CIS Ethics Committee and is member of the CIS Technical Committee on Games.

Dr. Greenwood is a member of the Tau Beta Pi and Eta Kappa Nu engineering honor societies and is a Registered Professional Electrical Engineer in the State of California.

Index

Printed in the United States
by Baker & Taylor Publisher Services